INGENIOUS
JIGS
&
SHOP ACCESSORIES

INGENIOUS
JIGS &
SHOP ACCESSORIES

Clever ideas for improving your shop and tools from *Fine Woodworking*

The Woodworker's Library

The Taunton Press

Front cover photo: **Vincent Laurence**
Back cover photos: **Strother Purdy (top left),**
Alec Waters (top right), Vincent Laurence (bottom)

Taunton
BOOKS & VIDEOS

for fellow enthusiasts

Printed in the United States of America
10 9 8 7 6 5 4 3 2 1

The Taunton Press, Inc., 63 South Main Street, PO Box 5506,
Newtown, CT 06470-5506
e-mail: tp@taunton.com

Distributed by Publishers Group West

Library of Congress Cataloging-in-Publication Data
Ingenious jigs & shop accessories : clever ideas for improving your shop and tools.
p.   cm.—(The Woodworker's Library)
ISBN 1-56158-301-4
1. Woodworking tools. 2. Woodwork—Equipment and supplies—Design and construction.
3. Jigs and fixtures—Design and construction.
I. Title: Ingenious jigs and shop accessories. II. Series.
TT186.I54   2000
684'.08—dc21   99-053022

About Your Safety
Working with wood is inherently dangerous. Using hand or power tools improperly or ignoring
standard safety practices can lead to permanent injury or even death. Don't try to perform operations you
learn about here (or elsewhere) unless you're certain they are safe for you. If something about an
operation doesn't feel right, don't do it. Look for another way. We want you to enjoy the craft,
so please keep safety foremost in your mind whenever you're working with wood.

"Necessity,
the mother of invention"

—GEORGE FARQUHAR

CONTENTS

INTRODUCTION

New woodworking tools are seldom ready to use right out of the box. New handplanes, for example, have to be sharpened before they'll cut well. Bigger, more expensive tools are no better. New bandsaws need complete assembly and tuning—they're really bandsaw kits. Even routers, which only need plugging in to work, are incomplete. A basic router is good for little more than edge treatments. It needs jigs, fixtures, and accessories to realize its versatility. And so it is with almost every woodworking tool: They're either rough or incomplete, and it's up to the buyer to make them useful.

Imagine if the same were true of automobiles. Would you buy a new truck if the wheels needed to be aligned before you could go anywhere? Or if you had to make your own headlights? Admittedly this is an unfair comparison. Few drivers want to be bothered with the mechanics of their cars; but this is not so with woodworkers and their tools. Woodworking tools are left rough and incomplete on purpose because manufacturers know we want them that way.

Inside every woodworker is a tinkerer and inventor trying to get out. If new tools were perfect, we'd be frustrated, not happier. We enjoy improving our tools, modifying and adapting them to specific tasks, even building them from scratch. Why? Because each successful improvement eliminates a problem that has dogged our work. Jigs and fixtures are like well-placed scratches on annoying itches. However, no two woodworkers will have itches in the same place. It would be surprising to find two woodworkers who could agree how to improve a router. Should the stock plastic base be replaced with a transparent acrylic, a plywood, or an MDF one?

This book is about unique and ingenious improvements to tools and the shop. The editors at The Taunton Press have brought together material from 38 articles in past issues of *Fine Woodworking* magazine. The authors are all woodworkers—some professional, some hobbyists. They tinkered with their tools and invented a wide range of jigs, fixtures, and accessories that help them work far more accurately, efficiently, and even enjoyably. Can a router be used to cut angled tenons and dadoes? How do you make true 90° crosscuts on a tablesaw? Is there any way to improve a portable planer's performance? Or is there a way to reduce shoulder and back pain while working? You'll find ideas for solutions to all of these questions and more to build your own jigs in your shop.

ONE

Materials for Jigs

A well-made jig will last a lifetime, repaying your investment in time and materials with every use. A poorly made one will only add to your frustration. Often the only difference between a good jig and a bad one is in the choice of materials. Most jigs guide tools and consequently are only as useful as they are accurate. Jigs should be made from stable and durable materials to stay on the money over long-term use and abuse.

A common mistake when building jigs is to pick random shop scrap. Convenience should not outweigh the construction requirements of a jig any more than it should for good furniture. Solid wood, though generally plentiful in the average workshop, isn't particularly stable. It warps when it grows in humidity and shrinks in aridity. For this reason, other materials such as plywood, medium-density fiberboard (MDF), and plastics are far better suited for jig-making than solid wood. Of course, in parts of jigs where stability is not an issue, random scrap will always be the best choice.

What follows is a short but essential section on MDF and plastics, two excellent materials for making jigs. MDF machines easily but is far more dense and stable than wood. It holds very crisp edges and won't wear as fast as wood. Plastics have similar attributes. They create less friction than wood, making them easy to move over workpieces or vice versa. Transparent plastics are ideal for safety guards because they don't block your view of the tool in action. And some are virtually unbreakable. Adding them to your material vocabulary will give you jigs that will be as accurate tomorrow as they were when you made them.

A WOODWORKER'S GUIDE TO MEDIUM-DENSITY FIBERBOARD

by Jim Hayden

A variety of different types of medium-density fiberboard (MDF) are available from nine companies that manufacture the material (above). Although it has been used in industry for more than 30 years, MDF has just recently become more available to consumers and small shops. The difference in density (right) is obvious when comparing MDF, left side, to common particleboard, right side.

Medium-density fiberboard, or MDF as it is more commonly known, is the newest of the furniture-quality wood composites. Because of its dense, uniform composition and flatness, it has surpassed plywood and particleboard as the sheet good of choice for fine work and more routine uses.

Pre-finished faces are flat as the slate on a pool table, which along with its dimensional stability makes it an excellent substrate for veneer. The edges machine well, with no chipout, and MDF accepts a full range of joinery and fasteners.

Museum quality MDF—These display cases at the Freer Gallery of the Smithsonian Institution show off some of MDF's versatility. The case on the left shows the crisp edge-holding ability of MDF in a painted piece. The case on the right makes use of walnut-veneered MDF with solid-wood moldings.

But if you have never seen a 4x8 or 5x8 sheet of MDF or have never even heard of MDF, you have plenty of company. MDF has been an industrial product for its entire 28-year history, with most shipments ear-marked for furniture factories and cabinet producers. Only recently has it become more available to retail consumers and small shops. Once you have some MDF in your shop, you may find, as I have, that it is also good stuff to make some of your jigs, fix-tures and templates.

Whether you use it for jigs or the sub-strate for fine veneered furniture, there are some special tricks and tips for using MDF. I'll share what I've learned from my own experience and from research done for the National Particleboard Association (NPA), which includes eight of the nine MDF companies, as well as from the reactions of woodworkers who regularly use MDF in the cabinet shop of the Arthur M. Sackler and Freer Galleries in the Smithsonian Institution (see the photos above).

Machining MDF

Because it's homogeneous (see the inset photo on the facing page), MDF machines better than plywood or particleboard, and even some natural woods. There are no lay-ers or chips, brittle edges, knots or grain. I routed all 15 types of MDF made in the United States, courtesy of the nine MDF companies. The boards share a sameness in meeting industry standards: They match in density and superb flatness. They differ because the trees harvested near the plants differ. The wood chips, shavings and saw-dust (or residuals) from the local sawmills and plywood mills are the raw material of MDF. Also, the companies use proprietary formulas, thus adding a few minor, and in some cases, a few major differences, such as formaldehyde content.

I've used two brands of MDF regularly during the last five years. I'm impressed with the consistently smooth surface of the sheets. MDF starts out as a low-density, 15-in.-thick slab 18 ft. long. A $^3/_4$-in. board is compressed at 800 lbs. pressure, then 50,000 lbs. pressure to almost final thickness. Sanders, in a series of grits, take over and finish off with 120- or 150-grit, sanding and burnishing to precisely $^3/_4$ in.

Sometimes I measure new sheets. I find their thickness to be scrupulously main-tained. However, extreme heat and humidity changes, such as daily changes encountered with outdoor storage, will cause a permanent thickness increase. But the thicker boards, $^3/_4$ in. and 1 in., will take some abuse in storage (i.e., stored on edge) and not warp.

Kerfs for a curve—
Woodworkers at the
Sackler Gallery
cabinet shop in the
Smithsonian
Institution kerf-bend
veneered MDF to
make a curved
museum bench.

Crisp profiles—
Medium-density
fiberboard excels in
maintaining sharp
edge profiles. Shown
here are, from left,
examples of cove,
roundover and
Roman ogee, all cre-
ated with a router.

Sawing

A 50-tooth combination blade is suggest-ed for rough-cutting large sections of MDF on the tablesaw. But I make so many things out of cutoff pieces that I go right to my finish-cut blade. That used to be a 60-tooth triple-chip. I loved that blade; with a pair of hold-downs and my pride and joy, shop-made, European-style adjustable splitter, a piece of MDF would slide down the fence and exit the blade with a new edge so smooth that I had to stroke it. Then I bought the other blade manufacturers recommended for MDF, a 60-tooth thin-kerf alternate top bevel (mine is a Freud TFLU88). It seemed to cut even cleaner than the triple-chip, and material moved more easily through the blade because of its semi-thin kerf (nominally .090-in.). Its teeth angles fit the NPA's specs for a blade to saw cleanly top and bottom surfaces of overlaid panels. They are a 15° hook, 15° top bevel and a 10° alternate face bevel.

I use 6-in. blade stiffeners for a slightly finer cut, and I made a zero-clearance insert to keep the dust down where it belongs. I'll talk more about MDF dust problems and solutions later.

My friends in the cabinet shop have good results using the tablesaw to kerf MDF sheets, so they can be bent into curved forms, as shown in the top photo at left.

Edge-shaping and routing

When I saw or rout an edge, rabbet or dado for joining, I get sharp edges with MDF. The edge surface looks and feels smooth. When rubbed counter to the cut direction, it feels slightly fuzzy or scratchy, depending on the brand of MDF, but the

piece is ready for glue-up and assembly. My contoured router cuts (cove, roundover and Roman ogee) are clean and smooth, with crisp edge profiles, as shown in the bottom photo at left. The edge surface is a little rougher than on straight cuts, but that disappears with normal light sanding for finishing. When routing or shaping, feed MDF about 25% slower than wood for maximum edge smoothness.

MDF does have its limits. Sharp protrud-ing contoured edges aren't a good choice. And being 10% urea-formaldehyde or other glue, MDF does wear down cutters faster than wood.

MDF sawdust is fine resin-coated par-ticles of wood dust, light enough to become airborne and settle on everything in sight. Building a router table/tablesaw extension with vacuum attachments has virtually eliminated floating dust and cut down on my set-up time (see the photo at left on the facing page). I use a high-quality dust mask, the Dustfoe 66, which I purchased from Highland Hardware (1045 N. Highland Ave. N.E., Atlanta, Ga. 30306; 800-241-6748) and have installed vacuum setups on all my machine tools.

I wanted the same low-dust environment for freehand routing, so I built an acrylic and MDF safety guard/vacuum hookup that bolts into T-nuts epoxied on the underside of my router base plate. It is almost 100% effective with MDF dust.

Successful sanding

Other advice to the contrary, don't sand an MDF panel before attaching an overlay. Scuff-sanding can cause a weaker glue bond. I just make sure my work table and the panel faces are nice and clean, then proceed.

Flat and contoured edges should be sand-ed before finishing to remove the nap. A belt sander is a good choice for flat edges, as is an abrasive wheel for contoured edges. Use a sequence of 100- to 150-grit, or 120- to 180-grit. It's a light sanding, not a dust raiser. Some shops prefer hand-sanding.

Sanding also is the process that can raise the most of MDF's extra-fine dust. At our cabinet shop in the Smithsonian, the helmet-type powered air-purifying respirator is used (see the photo at right on the facing page).

Shop-built attachments handle MDF's dust on the author's router table. An extension of his tablesaw table, the router table houses a shop vacuum and sound baffling, which makes for quiet, dust-free operation.

Dealing with dust—A worker in the Sackler Gallery cabinet shop at the Smithsonian Institution avoids the fine dust produced by MDF by wearing a Racal powered respirator while finish-sanding a display cabinet.

Joinery and glue choices

MDF machines and glues well, giving it literally a sharp edge over plywood and particleboard in joint-making. "We do a lot of case work with Medex MDF, using miter joints," explains a cabinetmaker at the Smithsonian. "I have to be careful handling the Medex edges. They are so sharp I've cut my fingers several times."

The furniture and cabinet industries use dowels extensively in MDF case work. Drawers are often made with $^7/_{16}$-in. dovetailed or rabbeted MDF sides. Independent furniture and cabinetmakers seem to be sold on biscuit joinery for MDF. If you use dowels, spiral and grooved dowels are recommended over plain dowels by eight of the nine MDF companies.

The joints that work well with MDF (as shown in the photo on p. 8) include: loose-tenons (spline-tenons), dovetails, sliding dovetails and finger joints. Spline or biscuit miters, lock miters, miter and rabbet joints, simple miters, rabbets, dadoes, and butt joints with biscuits or dowels also work well.

Adhesives

A high solids or gap-filling glue is ideal for MDF. I use modified (yellow) polyvinyl acetate (PVA) or Titebond II when assembly time permits. Otherwise, it's white PVA, the same as I'd use for wood. Contact cement, epoxy and urea resin work well when needed. A factory method is to use a hot-pressed rigid resin to bond an MDF core and hardwood veneer. Under low pressure at 250° for less than three minutes, the thermosetting glue doubles the panel's stiffness. I would rate urea resin and epoxy as the best thermosetting glue candidates, but it's best to make your own tests on scrap before the final glue-up.

Hardware and fasteners

A straight-shanked screw with deep, wide, sharp threads is best for MDF. I use Robertson square drive, particleboard and sheet-metal screws. (On the West Coast, a savvy MDF user recommends Twinfast particleboard screws.) The fine-threaded sheet metal or self-tapping screw is also good. And drywall screws make handy temporary holders for MDF projects.

Do not use tapered wood screws. Screw threads cut MDF fibers and resins. While regular wood springs back, MDF distorts. The distortion and tapered screw shape combine to make for poor fastening.

There's a limit to the screw size an edge will take without splitting (see the chart on p. 8). Use longer (not larger) screws in the edge for increased strength. More screws add strength, up to 4 in. apart.

A wide variety of joinery is possible with MDF. Examples shown here surrounding a routed dovetail joint are, clockwise from left, biscuits, dado, rabbet and dado, spline miter and biscuited miter. The background is a sheet of factory-veneered MDF.

Using screws in MDF

Screw size	Pilot hole	Minimum edge*
#6	3/32 in.	1/2 in.
#8	7/64 in.	5/8 in.
#10	1/8 in.	1 in.

*Minimum sheet thicknes for driving screws into an edge without splitting.

Drill pilot holes in the edge, so the board won't split, and drill them to the depth the screw will be inserted, plus about 1/4 in. It's also a good idea to drill pilot holes in the face plane. See the chart above for common screw sizes and correct pilot holes.

Screws in the face should be at least 1 in. away from corners, and edge screws should be 3 in. from corners. A slow drill speed or dull bit will burnish the pilot hole wall and cause crumbling. Run a sharp bit at high speed (3,000 rpm for industrial applications). You'll get a clean, accurate hole with top pull strength.

There's a "turns" trick to make sure you don't over-torque and strip the panel threads. A three-quarter turn past flush on the face is maximum torque. A three-eighths turn past flush on the edge is the limit there. Even with a properly sized and countersunk pilot hole, the screw will break away beyond these points.

My friends in the cabinet shop sometimes use pneumatically driven ring-shank coated nails or coated staples on glued joints to save clamping time. If you do that, be careful not to drive edge staples with their legs parallel to the surface, or you may get splitting.

Hinges

After trying all sorts of hinges, I found the best hinges for MDF attach face to face. When hinges are installed, MDF may "pyramid," or develop a bump around hinge screws. To prevent the pyramid and to ensure the hinge is flush, drill a partial countersink along with the pilot holes.

Laminating, veneering, and finishing

Good bonding strength, dimensional stability, flatness and other qualities previously mentioned make MDF an ideal substrate for numerous materials, including high-pressure laminates and veneers. Cross-banding is unnecessary with even the thinnest veneers.

Either veneers or paint can be used to finish edges. A painted edge may work well with laminated, veneered or, of course, painted face planes. It involves a typical edge-finishing process. Careful sanding is followed by one or two coats of sealer. Burnish smooth each coat of sealer before applying the final topcoat.

Quick-drying sanding sealers, auto-body primers and even white PVA glues diluted 20% can be used as edge sealers.

Sealing in formaldehyde may be a factor in finish selection. The level of formaldehyde in untreated MDF may remain above ambient levels for several years.

High-pressure laminates offer almost total sealing, matching factory applied thin and thick vinyls. After that comes alkyd oil primer and oil enamel paint combined, two coats of polyurethane, and latex-ammonia combined with two coats of latex wall paint. (The latex-ammonia types will raise the grain.) Ironically, the effective alkyd oil finishes contain formaldehyde, but it normally off-gases in two weeks.

Finishes that are less effective sealers include: oil base or lacquer sealer plus a top coat of varnish or lacquer; two coats of lacquer or oil primer; lacquer sanding sealer plus one or two coats clear lacquer; quick-drying lacquer sanding sealers; and shellac or varnish applied without a sealer.

Despite their other merits, finishes that will not effectively seal in formaldehyde in MDF include: two coats of regular latex paint, penetrating oil sealer, stains, waxes or linseed oil.

There are treated low-formaldehyde MDFs, such as Plum Creek, and formaldehyde-free brands, such as Medite II and Medex (exterior grade), to consider. I asked the Sackler and Freer cabinet shop supervisor, Cornell Evans, for his impressions of Medex. "Medex has no formaldehyde and is fire-rated. It is lighter and harder than (regular) MDF," he said. "It glues better and takes paint better. We use it for case work. It has sharp edges, is water repellent and is much less dusty (than other MDFs). There is no fine sawdust when cutting. We use $5/8$-in. Medex in place of $3/4$-in. MDF."

Finding and buying MDF

Standard MDF costs about 60% less than seven-ply birch plywood and about 40% more than particleboard. And formaldehyde-free, water-resistant Medex-type MDF is about triple the cost of particleboard, but it is still 15% less than top-quality birch plywood.

Locating and buying medium-density fiberboard is sometimes difficult because so much of it goes directly to industry. In the summer of 1990, several MDF companies began test-marketing their products around the country.

California is a big test market. The Medite Corp. based in Medford, Ore., is placing $3/4$-in. Medite in 20 Home Depot stores there. (Try Home Depot elsewhere for other brands.) J.E. Higgins, a chain of small lumberyards, has MDF in their yards in Los Angeles, San Francisco and Sacramento.

Some chains have MDF in selected stores across the country. The stores include: Handy Dandy, Channel and Lowes. Sequoia Supply in Columbia, Md., distributes Plum Creek MDF to lumberyards in parts of Maryland, Virginia and Pennsylvania.

If you know of mills that cater to woodworkers, call them. If you must special order, local independent dealers are your best bet. A chain that has particleboard but no MDF may be able to order some for you from its particleboard source, but you will pay top dollar.

Small commercial shops can buy from one of the 2,000 industrial wood products distributors in the United States. Would

the industrial distributor welcome me if I showed up as an individual to buy one or two sheets? Probably not. But my 275-member Washington Woodworkers Guild has an agreement with one to sell to all our members, large orders or small. (Guilds have buying power. We have price discounts from several stores, wholesalers and manufacturers.) Don't overlook the fact that some sellers are willing to deliver sheet material.

If all else fails, write one or more of the MDF companies listed at right, and tell them everything you went through and how badly you want their product. If you have equally interested friends or belong to a guild with a genuine interest in MDF among its members, mention that also. No one is promising you instant results, but many letters from many woodworkers do a market make.

SOURCES OF SUPPLY

The following companies manufacture medium-density fiberboard. Contact them for the name of a distributor near you.

GEORGIA-PACIFIC CORP.

Holly Hill, 133 Peachtree St., N.E., PO Box 105605, Atlanta, GA 30348

INTERNATIONAL PAPER, MASONITE DIVISION

Spring Hope Plant, Highway 64 and County Road 1306, PO Box 369, Spring Hope, NC 27882; Marion Plant, Highway 301, PO Box 8, Sellers, SC 29592

LOUISIANA-PACIFIC CORP.

Eufaula Mill, Route 3, Box 22, Clayton, AL 36016; Oroville Mill, PO Box 158, Samoa, CA 95564

MEDITE CORP.

PO Box 4040, Medford, OR 97501; PO Drawer 1427, Las Vegas, NM 87701

NORBORD INDUSTRIES, INC.

PO Box 26, Deposit, NY 13754

PLUM CREEK MANUFACTURING, L.P.

PO Box 160, Columbia Falls, MT 59912

SIERRAPINE LTD.

product sales by Timber Products Sales Co., PO Box 269, Springfield, OR 97477

WEYERHAEUSER CO.

PO Box 290, Moncure, NC 27559

WILLAMETTE INDUSTRIES, INC.

Bennettsville, PO Box 636, Bennettsville, SC 29512; Malvern, PO Drawer 190, Malvern, AR 72104

PLASTICS FOR WORKSHOP JIGS

by Jeff Kurka

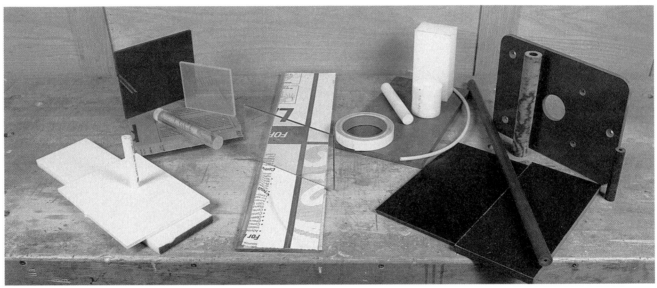

Myriad plastics help woodworkers make better jigs, fixtures and shop-built tools. From left: Slippery acetals make good bearing surfaces; rigid and clear acrylics are handy for templates; the impact resistance of polycarbonates protects you as machine guards; polyethylenes are tough and slick and great for sliding parts; and rigid and strong phenolics have no peer for table-mounting routers.

What's an article about cold, hard plastics doing in a woodworking publication? Plastics and wood may have very little in common, but it's the differences that complement each other. Primarily used for jigs and fixtures, as shown in the photos above and on the facing page, these materials enable woodworkers to develop more professional and more durable shop tools and apparatus. Plastics have a high strength-to-weight ratio, they are relatively hard and smooth, and they will far outlast wood when used as bearing or guide surfaces. Most plastics are extremely stable, being unaffected by moisture and only slightly affected by extreme temperature changes (about 0.0001 in. per degree C). Many plastics are easily worked on regular woodworking equipment and can be fastened chemically or mechanically. Some plastics even can be heated and reshaped around simple forms or patterns. And because plastics are available in a wide range of sizes (thicknesses, lengths and widths),

shapes (round or square rods, sheets, blocks or tubes), colors and opacities, finding the right plastic is relatively easy.

There are five types of plastics that have found uses in my shop; acetals, acrylics, polycarbonates, polyethylenes and phenolics. Although the machinery and cutting tools to work these plastics may be familiar to you, the reaction of the plastics to machining may not be; therefore, it's important to understand the characteristics of each plastic to ensure quality parts and safe operating procedures.

Acetals

Acetals are part of the thermoplastic family (synthetic resins that soften or fuse when heated and harden again when cooled). Developed by Du Pont under the product name of Delrin, acetals are the toughest and most resilient of the thermoplastics.

High in tensile and compression strength, acetals are dimensionally stable and also

Plastics in the shop—Although this is a woodworking shop, plastics play a vital role, from holding workpieces in place, guarding cutting tools and guiding crosscut boxes to mounting routers and providing work surfaces. No shop would be complete without these handy materials that can make your shop-built tools more professional.

slippery. Because they are almost impervious to solvents, they are impossible to glue together, but they take and hold screws much like metal.

Acetals' slippery nature and durability make them ideal for rollers, bearings, bushings or guide bars to ride in a tablesaw's miter slots.

Acrylics

Another kind of thermoplastic, acrylics are more familiar by their trade names, Plexiglas and Lucite. Although available in a variety of grades, I've found the standard grade to be sufficient for most shop applications. Acrylics are dimensionally stable and extremely rigid but brittle. They have moderate-to-high impact strength, which can be increased by bending and forming them into curved shapes.

Acrylics have high optical clarity and transmit 3% to 5% more light than standard plate glass, but they scratch easily.

Acrylics can be chemically bonded by solvents or polymerized cements and mechanically fastened by a variety of means, but care must be taken when tapping or threading because the material is brittle.

Acrylics' optical clarity and rigidity make them well-suited for router baseplates, machine guards and parts templates.

Polycarbonates
Polycarbonates, such as Lexan, are often confused with acrylics and actually have some similarities; however, there are many notable differences. Polycarbonates are virtually unbreakable with an impact resistance 250 times greater than glass and 30 times greater than acrylic. Optical clarity is high, but dimensional stability is only fair due to their tendency to absorb moisture. Polycarbonates are approximately 20% more flexible than acrylics and may sag under a constant load (such as a table-mounted router).

Polycarbonates can be fastened chemically with solvents or mechanically with a variety of fasteners.

Shop applications are similar to those of acrylics, but the greater impact resistance of polycarbonates makes them particularly suited for machine guards.

Ultra-high molecular-weight plastic
Ultra-high molecular-weight (UHMW) plastic is a member of the polyethylene family and has superior physical and mechanical strength. UHMW is more resistant to abrasion and chemicals than stainless steel, it's slipperier than plate glass and it's relatively unaffected by moisture. Although UHMW can experience an expansion or contraction rate of almost $1/2$ in. per 8 ft. at a 50° temperature swing, it's acceptably stable when mechanically fastened in place. UHMW also exhibits a moderate-to-high impact resistance. Although UHMW is a thermoplastic and will start to soften at around 170°F, it cannot be heated and reshaped.

UHMW can be machined with many traditional woodworking machines and tools and holds threads well for mechanical fastening. It cuts well on the tablesaw, but the shavings are fine and feathery and cling annoyingly to everything. UHMW can even be thickness-planed with excellent results.

UHMW also is available as tape in a wide range of thicknesses and widths. As solid stock or tape, UHMW is great for adding slippery, tough surfaces to sliding parts on jigs or fixtures, as drawer glides or as skid plates on cabinetry.

Phenolics
Phenolics, also known by the trade name Micarta, are of the thermosetting family and thus become permanently rigid when heated or cured. Phenolic resins, silicones, melamines and epoxy resins are formulated in various combinations and then reinforced with paper, canvas, linen and glass to create the various grades available. Applications of the grades (too numerous to mention here) are broken down into either electrical or mechanical. I've found a general paper/mechanical grade to be more than adequate for my woodworking applications, but you might consider a cloth/mechanical grade if your application requires close tolerances on small parts such as gears or pinions.

Phenolics are among the hardest of the plastics and rate high in impact and compressive strength and dimensional stability. They are extremely rigid and will bend or stretch very little before breaking; however, thickness tolerances can vary as much as 0.015 in. in a 36-in. span.

Phenolics can be fastened using two-part epoxies or many mechanical fasteners. They tap fairly well, and the threads will stand up to normal use. For repeated use, you should use brass inserts.

Phenolic's strength, stability and rigidity make it a good metal substitute in many applications and an excellent choice for table-mounting routers and router subbases.

Machining plastics
The plastics that I've been discussing can be machined on a variety of woodworking equipment: tablesaws, bandsaws, scrollsaws, drill presses, sanders, pin routers, router tables and, in a few cases, jointers and planers. Portable power tools and hand tools round out the list.

In general, tools should be operated at high speeds with moderate feed rates. The rpm of most power tools, such as tablesaws and routers, is acceptable for machining

Plastics for jigs and fixtures—A shop full of accessories can be built with plastics. Clockwise from the bench vise: a mortise jig, a router subbase, a dovetail fixture, a featherboard, a hinge jig, a router-bit guard, hold-down clamps and a router-bit setting gauge.

plastics. Drill press rpms can and should be varied to match their particular tooling or operation and will be dealt with when I discuss the drilling techniques.

Heat buildup is the biggest enemy to a good cut. Coolants, such as water or air jets, can be used to reduce heat, although for short runs, they are rarely necessary with sharp tools and proper feed rates.

Feed rates vary slightly depending on the material and the thickness of material being cut. A good starting point is 2 in. to 3 in. per second. Watch the point of the cut, and look at the resulting finish. Too slow and the plastic becomes gummy and sticks to the blade, especially with thermoplastics like acrylics and polycarbonates; too fast usually results in crazing and chipping of the plastic because the cutting tool loads up. With plastics such as acrylic, polycarbonate, acetals or polyethylene, the cutting tool tends to grab the material, and your edge finish will be poor. The cut will sound raspy rather than smooth and clean. On a hard material like phenolic, the plastic may actually climb the sawblade and kick back, especially with a tablesaw. Listen and look at what the tools are telling you, and with a little practice, you'll soon develop the proper feed rate for each type of plastic.

Safe working practices

Some of the acrylics, polycarbonates and a few phenolics can release mild concentrations of chemical vapors while being machined. Prolonged exposure can cause eye and respiratory irritation, headache and nausea. Because exposure is limited, this shouldn't be a problem, but there are precautions you can take.

Heat buildup when machining releases vapors, so air jets directed at the point of machining help cool the plastic and slow the release of vapors. Also, air jets help cool the cutting tool and blow away chips, shavings and vapors. Dust collection helps stop the plastic dust from becoming airborne. Finally, as is the case when cutting wood, adequate shop ventilation, suitable dust masks, machine guards and safety glasses will go a long way toward making your plastics experience a pleasant one.

Rough-cutting

I've found it best to rough-cut plastics slightly oversized and then machine to the finish size, usually with a template and router. I rough out straight cuts on the tablesaw and curved or irregularly shaped cuts on the bandsaw.

The ideal tablesaw blade for all plastics (except phenolics) is an 80-tooth, carbide, triple-chip design, which will give you the smoothest cut and the longest blade life. An 80-tooth, carbide, alternate top bevel (ATB) blade, either crosscut or plywood style, is a good alternative for the occasional user. I raise the blade about $^1/_2$ in. above the plastic to help reduce the heat.

Phenolics present special problems because they are so hard. An 80-tooth, carbide, ATB crosscut blade is okay for occasional use, but repeated cuttings will quickly dull this blade. A better choice is an 80-tooth, carbide, triple-chip design with negative hook: It's not as aggressive as standard wooden blades. Raise the blade only about $^1/_8$ in. above the material, and feed as rapidly as possible without forcing the phenolic through the blade.

Bandsawing acrylics, polycarbonates, acetals and polyethylene up to $^1/_2$ in. thick is best done with a $^1/_8$-in.- to $^3/_8$-in.-wide blade with 10 to 14 teeth per inch (t.p.i.). I use the smallest blade that still gives a good cut because the wider the blade, the greater the heat buildup. Periodically, clean your bandsaw's wheels; plastic sawdust quickly builds up on the rubber wheels and can affect blade tracking. The same rules apply to bandsawing phenolics except you'll get a better cut from a blade with 4 to 7 t.p.i.

Sabersaws also can rough-cut plastics, but you must support the material as closely as possible to the cut to prevent cracking. A $^1/_4$-in.-wide, 10- to 14-t.p.i. blade should give good results.

Drilling acrylics or polycarbonates—A regular drill bit run at 300 rpm will work in acrylics and polycarbonates if you back it out frequently to clear chips (left hole). For higher speed production work, a special bit with a 60° point works best (right hole).

Routing

Just as when routing wood, it's best to make your rough cut within $^1/_{16}$ in. to $^1/_8$ in. of the finish line, and then use the router either freehand or in a table for a final trimming cut. Although router bits can safely and effectively cut up to 25% of the diameter of the router bit, the less you trim, the smoother the finish will be.

I prefer a high-speed-steel spiral bit for edge-trimming acrylic, polycarbonate, acetals and polyethylene; its shearing cut yields the cleanest edge. Two-fluted, carbide-tipped bits, both straight and spiral, with and without bearings, also work well with these plastics. Spiral end mills can be used when cutting slots.

As a general rule, use the largest diameter bit that your work will allow. I try to use at least a $^1/_2$-in.-dia. bit whenever possible to reduce chatter. For smaller bits, slow the feed rate to allow adequate chip clearance.

The hardness of phenolics restricts you to carbide-tipped bits. Be extra careful when using a carbide bit smaller than $^3/_8$ in. dia. Small bits are commonly solid carbide, which is brittle and will snap easily.

When trimming phenolics, rough-cut as close to finished size as possible. A heavy cut in phenolic will cause the bit to grab with enough force to tear the workpiece from your hands. When template-routing, start your cut in the middle of a straight section, feed slowly with a good grip and you'll get good results.

For interior cutouts in any plastic, it's best to cut away the waste just as for an outside cut. I don't recommend plunge-cutting because the bit tends to grab the waste piece and throw it out violently. Slots can be cut by drilling a start and stop hole and then routing through the plastic in a series of passes.

I cut sliding dovetails or T-slots by first wasting as much material as possible with a straight bit or with the tablesaw. Then the dovetail or T-slot bit is making only a cleanup cut, which reduces heat buildup and bit loading and leaves a cleaner slot.

Drilling

When drilling, remember excess speed causes excess heat. Follow the recommended rpm for the type of drill bit you're using. As a general rule, thicker plastics will require slightly slower speeds.

High-speed-steel twist bits will work on all of the plastics in this article. Drill press speeds of 700 rpm to 1,000 rpm deliver good results when drilling acetals and polyethylenes, but 300 rpm is best for acrylics, polycarbonates and phenolics.

A bit with a 60° angled point (see the photo on the facing page) produces better results on acrylics and polycarbonates, and you can bump the rpms to 1,500-2,000. Carbide-tipped bits work much better when drilling phenolic.

For clean holes with no chipping, use a wooden backing board. To protect your hands, always clamp the workpiece down. Use a slow, even feed rate, and back out of the hole often to reduce heat buildup and to clean the bit.

Sanding

Machine sanding with 100-grit to 120-grit paper is okay for roughing plastics to size, but remember to use a light touch and back away from the abrasive often to avoid distorting the edge.

After machine-sanding, hand-sand, starting with 280-grit paper and ending with 600-grit to leave a smooth edge without any sanding marks. If desired, edges could be polished with rouge and buffed.

Gluing

Acrylics and polycarbonates are easily glued together with a methylene chloride-based solvent that breaks down the plastic and fuses the pieces together. If your joints are tight-fitting and clean and you follow the manufacturer's instructions, you can get almost invisible joints.

Acetals, polyethylene and phenolics are not easily glued, and stronger bonds are possible mechanically.

Tapping and threading

All of the plastics can be successfully tapped and threaded, but take extra care when working with acrylics because they are a little brittle. Use a national coarse thread (it holds better than a finer thread). Wax lubricant on the taps and dies aids the cutting operation and leaves the threads cleaner and more transparent. Backing out of the cut often will clear the chips and improve the quality of the threads. Finally, if you are tapping a hole drilled into the edge of the plastic, clamp the edge in the jaws of a vise or hand-screw clamp to reduce the chance of cracking.

TWO

Workbenches and Bench Fixtures

What's the difference between a jig and a fixture? An easy question, one would hope. Both are accessories to tools that guide or assist in some action. On the surface, the only difference is that jigs move (a jig is also a dance) and fixtures don't (they're "fixed"). These definitions seem logical, but don't push them too far or you might ask why dovetail jigs aren't called dovetail fixtures. After all, the router does the moving and the jig is, well, fixed. In fact, it's difficult to think of a tool you'd call a "fixture." It's probably easier to just drop fixture from the shop vocabulary and simply stick with "jigs."

The difficult question comes when you push the definition of jigs because it's often hard to tell them apart from tools. Consider this: Is a workbench a jig or a tool? On one hand, it's a major shop item, and "jig" seems such an inadequate word for it. On the other hand, the main use of a bench is very jig-like: to hold workpieces while you work on them with other tools and act as a reference. Benches fit the wider definition of a jig.

In the end, the difference between a jig and a tool is a matter of perception. Look hard enough, and you'll always find a jig somewhere in a tool and vice versa. That is what some of the authors in this section have done. They have found a few of the many jigs hidden in the common workbench and brought them out. When shrunk down, a workbench becomes a low assembly bench. When you add a vacuum cleaner to a bench, you get a vacuum hold-down table. This section also explores more traditional bench and shop accessories, such as a shooting board, an improved sawhorse, and two tool-and-clamp storage solutions. Not to detract from these authors' ingenuity but rather to enhance it, these articles show that the most useful jig is sometimes the simplest.

BUILD A BETTER SAWHORSE

by Voicu Marian

The well-built sawhorse

Optimal dimensions for these horses depend on the function for which they're intended and on individual height and preference. For someone of average height, 32-in. horses make a good base for an auxiliary workbench, and 24-in. horses are about right for an assembly and finishing platform.

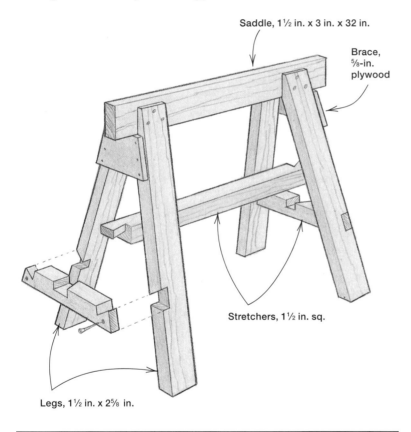

Saddle, 1½ in. x 3 in. x 32 in.

Brace, ⅝-in. plywood

Stretchers, 1½ in. sq.

Legs, 1½ in. x 2⅝ in.

I made my first pair of these sawhorses a few years back while remodeling my house because it was uncomfortable working stooped down on the floor. With a hollow core door on top, I had a fairly sturdy workbench that could be moved easily from one room to the next. After finishing up in the house, I took them back to the shop, and that's where they've proven their value.

My workbench always seems to be cluttered with tools. Before I made these horses, I often used the tablesaw as an auxiliary bench. That worked well as long as I didn't need to cut anything.

Now I have a second workbench: A pair of these sawhorses provides a strong, stable base; a couple of thick, heavy planks atop them form a perfectly serviceable benchtop; and a plank across the stretchers makes a good shelf for bench planes and other larger tools that normally clutter a bench surface. I clamp horses and planks together for stability and use C-clamps and bar or pipe clamps in lieu of vises, dogs and bench stops. When I'm finished with the bench, it disassembles and stores easily.

What makes these horses different from most, though, is the joinery. I first saw this half-lap, half-dovetail joint (see the drawing at left) used by an old carpenter when I was growing up in Romania. It's a strong joint, not too finicky to cut—especially in softwood. The joint gives these horses greater strength and rigidity, a much longer life and, as a bonus, a nice look. Also, the practice you gain in laying out and cutting the joinery in construction lumber will transfer to the fine work you do in hardwoods.

Construction sequence

I dimension all my stock first and then bevel all the edges with a block plane. To ease assembly and ensure consistency, I nail together a quick, simple set-up jig, consisting of three pieces of scrapwood on a plywood base (see the photo at right).

I determine the angle of the legs by eye rather than by using any mathematical formula. I hold two legs upright and adjust their spread until it looks right. Checking with a protractor for future reference, I read 35°.

I cut the notches at the top of the legs for the saddle first, space the legs with a block the same size as the saddle and then lay out the short end stretchers. I lay out and cut the half-lap first, scribing from the insides and outsides of the legs. I mark out the dovetails on the top side of the stretcher at 8°, cut them and scribe around them with a sharp pencil onto the legs. I cut and chisel out the leg to receive the stretcher. When the joint is assembled, leg and stretcher should be flush.

With all four end assemblies complete, I stand up a pair at a time and install the saddle, leaving a 4-in. overhang at each end. This provides a wider support for the boards I use as a benchtop as well as clearance for my feet. For now, one screw holds it together. Next I adjust the sawhorse so it's square to the surface it's standing on. Then I place the long stretcher across the short ones. I center it and mark it for length and for the shoulder of the half-lap.

A simple, nailed-together jig speeds layout and ensures consistency from horse to horse. Here, the author scribes around the end stretcher dovetail to cut out its mortise in the leg.

The rest of the process is the same as for the short stretchers, except I put the dovetails on opposite sides at each end. I do this more for aesthetics than for any structural reason. I glue the long stretcher in place and screw it from below. I then put two more screws on each side of the legs for a total of three screws into the saddle at the top of each leg.

The last thing I do is cut the tips of the feet, so they don't rock. To mark them, I lay the sole of my square flat on its side, scribe around each foot and then saw them off.

LOW ASSEMBLY BENCH

by Bill Nyberg

Make clamping easy—Two vises that can be adjusted independently hold even irregular shapes securely.

The open space at the center of the bench allows clamping pressure to be applied anywhere.

My father learned woodworking in Sweden, and when he came to this country, he got a job building reproduction Early American furniture. The shop had been in operation since the late 1700s, and like those who worked before him, my father was assigned a huge bench with many drawers. He stored his tools and ate his lunch at the bench, but much of his actual work took place nearby on a low table he called "the platform."

When I inherited his big bench, I also found myself doing most of my work at a low platform improvised from sawhorses and planks. I have bad shoulders and the occasional sore back, so using a full-height bench is difficult and unproductive. I needed a bench that suited the way I really work, so I built a low platform that incorporates some features of a traditional full-sized bench.

A clamping machine

My low platform bench is made for clamping (see the photos at left). The edges overhang enough for clamps to get a good grip anywhere along the length of the bench. A 4-in.-wide space down the middle increases the clamping options.

This platform bench has four tail vises made from Pony No. 53 double-pipe clamps, which can be used by themselves or in combination with a row of dogs on the centerline between the screws, as the drawing shows. Unlike most bench arrangements, with a single row of dogs along one edge, this one doesn't twist or buckle the piece. I can use each vise singly or with the others because the pipes are pinned into the benchtops at each end with $^1/_4$-in. by 2-in. roll pins. Without the pins, the pipes would slide through the bench when tightening one end.

Rather than using traditional square bench dogs, I bored $^3/_4$-in. holes for a variety of manufactured dog fixtures or shopmade dowel dogs (see the drawing on the facing page).

Building the benchtops

The bench is made from eight straight, clear 8-ft. 2x4s that I had kept in the shop for a few months to dry. I jointed the edges then ran each of the boards through the planer until the radiused corners were square.

Building the legs and base according to the dimensions on the drawing is straightforward. The leg braces are resawn 2x4s, about $^{11}/_{16}$ in. by $3^3/_8$ in. The only point to note is the dovetail connecting the beams to the legs. Because of the orientation of the beams and legs, the dovetail is only $1^1/_2$ in. at its widest point, but it's $3^1/_2$ in. from top to bottom. I tilted the tablesaw blade to cut the tails on the beam and cut the pins on the legs in the bandsaw. Almost any method would work to join the beam to the leg; my first version of the bench used a bolted slip joint.

The pipes run through the tops

The tops are made in two sections and glued up with the pipes and vises in place. The upper sections are made of three boards and the lower section from two. I edge-glued them with alternating growth rings to eliminate cupping. I cut $^7/_8$-in. grooves lengthwise in the top face of the bottom section to accommodate the pipes.

The tops are held to each beam with a single lag screw, which allows seasonal movement. To lock the tops into the base, I cut dadoes on the lower faces of the bottom sections to fit over the beams.

Assembling the double-pipe clamps

The double-pipe clamps are sold with a tail stop and a screw head. I set aside the tail-stop ends and used only the screw heads. Threading on the vise at one end of the pipe will unscrew the vise at the other end. So I had a plumber cut the threads twice as long on one end of each of the four pipes. I threaded the first vise all the way onto the end with double-long threads so that it was twice as far on the pipe as it needed to go. By the time the second vise was in place, the first one had unscrewed itself to the correct location.

Keeping ends flush when gluing

Before the pipes are installed in the grooves, I cut all the bench pieces to length. Once the tops are glued up, the pipes and vises are in the way, so it's hard to trim up ends that aren't flush. For flush ends, I aligned the pieces with dowel pins between top and bottom. I applied the glue and clamped the top and bottom sections together with the dowels in place. After the glue was dry, I drilled for the roll pins from the bottom so they wouldn't show.

A low bench made for clamping _____

This bench is 24 in. high, a convenient height for working on many projects. The benchtops are 42½ in. long, which gives more than 4 ft. between the jaws. At about 70 lbs., the bench is light enough to move around yet heavy enough for stability.

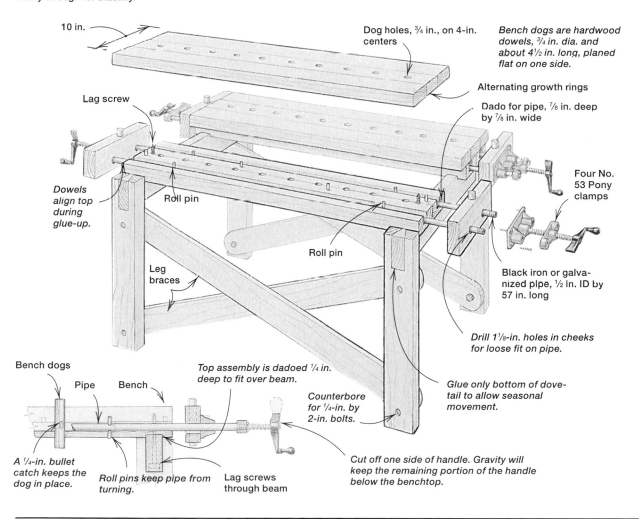

10 in.

Dog holes, ¾ in., on 4-in. centers

Bench dogs are hardwood dowels, ¾ in. dia. and about 4½ in. long, planed flat on one side.

Lag screw

Alternating growth rings

Dado for pipe, ⅞ in. deep by ⅞ in. wide

Dowels align top during glue-up.

Roll pin

Four No. 53 Pony clamps

Leg braces

Roll pin

Black iron or galvanized pipe, ½ in. ID by 57 in. long

Drill 1⅛-in. holes in cheeks for loose fit on pipe.

Bench dogs

Pipe   Bench

Top assembly is dadoed ¼ in. deep to fit over beam.

Counterbore for ¼-in. by 2-in. bolts.

Glue only bottom of dovetail to allow seasonal movement.

A ¼-in. bullet catch keeps the dog in place.

Roll pins keep pipe from turning.

Lag screws through beam

Cut off one side of handle. Gravity will keep the remaining portion of the handle below the benchtop.

DEDICATED BENCH FOR DOVETAILING LARGE CARCASES

by Charles Durham Jr.

I made my first dovetailed carcase with wide pine boards salvaged from the original kitchen in my first house. Dry, flat and wide, those boards became a wonderful blanket chest. Since then, much of the lumber I've used on large-carcase projects has been less than ideal. Wide, flat and dry are more the exceptions than the rule, whether you use naturally wide boards or glue narrower stock to width. When wide boards are cupped, twisted or both—even a little—making dovetails that fit well is tough. Yet accurately fitted and squared dovetail corners are crucial to the success of large projects like blanket chests, highboy tops and slant-front desks.

The other problem with large-carcase projects is the glue-up. Even if you've cut good, accurate dovetails, gluing and clamping big boards can be a real headache or, worse, result in a flawed project—especially if you work alone, as I usually do. Having the pipe clamp I just tightened fall off and dent the carcase as I tighten the next clamp is just one more hassle than I need.

I solved both problems by building two assemblies: a dovetailer's bench to hold the boards flat, secure and indexed for accurate layout and cutting (see the drawing on the facing page and the photos on p. 24) and a carcase-press clamping system to help me close wide joints with uniform pressure, without having to wrestle an armload of clamps (see the photo on p. 25). Material for both is available at any good lumberyard, and you'll find all the hardware you need either at your local hardware store or through mail order. Total cost for materials was about $300, with lumber being the most expensive item. By substituting construction lumber for the hard maple I used, you could halve that amount.

Dovetailer's bench

The problem with laying out and cutting dovetails on a typical cabinetmaker's bench is that most benches are about 32 in. off the floor, which constrains you to narrower carcase work. To do bigger jobs on an ordinary bench, you have to jury-rig a support and clamp system to hold things flat and steady at the right height while you mark, saw and chop. My bench is a large, elevated clamping device that lets me overcome warp on wide boards, allowing me to dovetail the largest boards with ease and precision. The bench's working surface is at elbow height: 42 in. off the floor, which is long enough for the longest pin member I'm likely to encounter.

The deepest carcase I would ever dovetail is about 25 in. So I added space for the clamp heads (see the drawing) to establish the benchtop's width of 28 in. A 72-in.-wide breakfront was the longest project on which I saw myself using the bench, so I decided to make it a bit more than half that length (48 in.) to keep that breakfront's top and bottom from falling off.

Dovetailer's bench

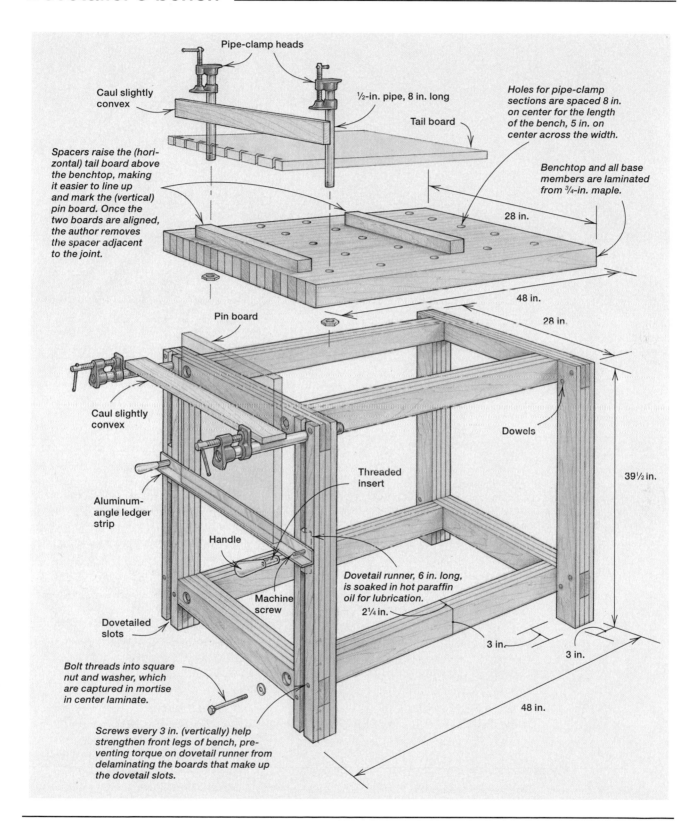

Pipe-clamp heads

Caul slightly convex

½-in. pipe, 8 in. long

Tail board

Holes for pipe-clamp sections are spaced 8 in. on center for the length of the bench, 5 in. on center across the width.

Spacers raise the (horizontal) tail board above the benchtop, making it easier to line up and mark the (vertical) pin board. Once the two boards are aligned, the author removes the spacer adjacent to the joint.

Benchtop and all base members are laminated from ¾-in. maple.

28 in.

48 in.

28 in.

Pin board

Caul slightly convex

Dowels

Aluminum-angle ledger strip

Threaded insert

39½ in.

Handle

Machine screw

Dovetail runner, 6 in. long, is soaked in hot paraffin oil for lubrication.

2¼ in.

Dovetailed slots

3 in.

3 in.

Bolt threads into square nut and washer, which are captured in mortise in center laminate.

48 in.

Screws every 3 in. (vertically) help strengthen front legs of bench, preventing torque on dovetail runner from delaminating the boards that make up the dovetail slots.

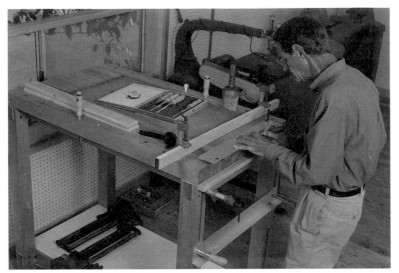

Marking tails—The author uses an aluminum template to mark out the tails on the side board of what will be a mahogany blanket chest. The short sections of pipe clamp at the front of the dovetailer's bench ensure the board remains flat for an accurate layout.

Marking pins from tails is more certain with a chisel than with a knife because there's no danger of the chisel following the grain. It's important, though, to make sure the chisel is absolutely perpendicular to the surface of the board you're dovetailing.

I use pipe-clamp heads to hold boards in place (see the photo above) and cauls extending across the bench's width to take out any warp in either board. An aluminum angle that raises, lowers and locks with a twist of the wooden handles serves as a ledger strip for the pin member (see the drawing on p. 23).

I cut dovetails in a fairly conventional manner, but with a couple of twists. I lay out the tails first, using a sheet aluminum template I made for the purpose. Then I saw to the line with a Bosch barrel-grip jigsaw and chop the waste out on my dovetailer's bench. The jigsaw is so much faster and is at least as accurate (probably more so) as cutting with a backsaw. I mark the pins from the tails, aligning the tail board on the benchtop with the pin board on the aluminum ledger, using a chisel and mallet to transfer lines (see the photo above right). A light, clean rap ensures a sharp line with no chance of following the grain, which can happen when marking with a knife. Again, I use the jigsaw, this time with its base set at approximately 14° (from a bevel-square set

on the tail board) to cut to the line and then chop out the waste on the bench. The fit I get with this system is nearly perfect.

Carcase press

My carcase press will close any size project I'll ever build and will do it in much less time than it takes with loose clamps. With the time saved, I can close the joints correctly before the glue grabs. The only fixed dimension is its internal working height—enough to take those 25-in. boards I produced on the bench. The carcase press consists of a pair of clamping frames made of maple laminations and pre-punched, galvanized steel strapping. The head member of each is fixed and has veneer-press screws mounted to it. (Veneer-press screws are available from Constantine, 2050 Eastchester Rd., Bronx, N.Y. 10461; 800-223-8087.) A foot member moves along the galvanized strapping to accommodate carcases of various widths. The clamping frames can themselves be positioned as near or far from one another as need be (see the photo on the facing page).

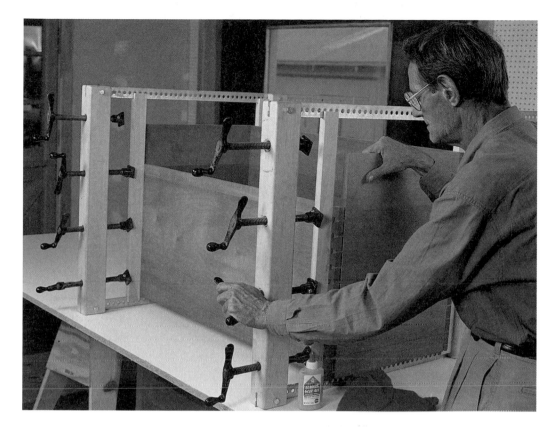

Clamping up a large carcase is much easier with the author's carcase-press clamping system than with ordinary pipe or bar clamps. The press consists of two units, each of which is made of four veneer-press screws, a couple of lengths of heavy metal strapping, a few board feet of hardwood and a handful of nuts, bolts and washers.

At each end of the maple laminations, I made a sawcut precisely as deep as the strapping is wide and drilled holes for the bolts that connect the wooden end pieces to the metal strapping. The straps I use are 60 in. long, but they're available in virtually any length. Smaller wooden cauls ride on the strapping to transfer the clamping force from the press screws to the carcase. Ideally, the clamping force should bear directly on the corner of the carcase, but I find that placing the force just inside the joint, right on the baseline, works just as well. With the 8-in. press screws and this setup, there's a range of about 4 in., fully opened to fully closed.

The elimination of loose clamps is the major benefit provided by the carcase press. Instead of watching and worrying about clamps falling off, I can monitor the joint. But there's another advantage. Quite often, clamping a project together forces it out of square in one plane or another. With loose clamps, the unending adjustment required

to restore squareness can be maddening. None of that has been necessary since I began using the carcase press.

Moreover, when using loose clamps, if a carcase winds (so that diagonal corners are high), there's nothing you can do with ordinary clamps. With the carcase press, I just wedge shims between press and carcase in the high corners, and it's flat again.

In using the carcase press, I work at table height on a sheet of laminate-covered particleboard. Because the bottoms of both clamping frames that make up the press are square, they stand upright on their own, making it easy to slide the carcase into the press. I get the joints just started outside the press and then place it inside and dry-assemble the carcase. Only after checking to see that everything's going to close up properly do I apply glue and clamp the carcase for good.

MAKING A T-SLOT TRACK

by Sandor Nagyszalanczy

One of the handiest methods of joining jig parts that must adjust is to use a T-track and T-bolt fasteners. A T-track is a useful way to mount fences, stops, hold-down clamps or to attach auxiliary tables and more. You can rout a T-slot into any solid wood, plywood or medium-density fiberboard (MDF) surface with a special T-slot bit (available from Woodhaven, 800-344-6657, or The Woodworkers' Store, 800-279-4441). The Woodhaven bit requires a $^1/_4$-in.- or $^5/_{16}$-in.-dia. straight bit and cuts a T-slot best suited to $^1/_4$-in.-dia. T-bolts or toilet bolts. The Woodworkers' Store T-slot bit needs a $^5/_{16}$-in.- or $^3/_8$-in.-dia. groove and is best for $^5/_{16}$-in.-dia. T-bolts.

The T-track slot is cut in two passes. The first pass, with a straight bit, makes a plain groove as long as the desired track length. The second pass is taken with the special bit that cuts the T-slot at the bottom of the groove (see the drawing below). For applications where a more durable slot is needed, The Woodworkers' Store offers a pressed-steel track that fits $^5/_{16}$-in.-dia. T-bolts. The track, which comes in lengths of 40 in. and 60 in., can be cut with a hacksaw and is designed to be epoxied into a $^{13}/_{16}$-in.-wide, $^{13}/_{32}$-in.-deep slot.

To attach parts or devices to a T-track, use T-bolts or T-slot nuts that ride in the track. T-bolts are available in $^1/_4$-in. and $^5/_{16}$-in. sizes and a variety of lengths. Standard toilet bolts (found in hardware stores) can also be used but not in all T-tracks. T-bolts may be secured with a regular nut, wing nut or hand knob. Standard carriage bolts can be used in T-tracks, but the depth of the T must be increased with the T-slot bit to clear the head. Carriage bolts can't take as much torque as T-bolts can without stripping the edges of the slot. T-slot nuts (available from Woodhaven) fit several different screw-thread sizes, from 10-24 to $^3/_8$ in. These are secured using a machine screw, a bolt or a studded hand screw.

Routing a T-slot

1. Rout a straight slot $^1/_4$ in. to $^3/_8$ in. wide to full depth of T-slot.

2. Use special T-slot bit to complete inverted-T-shaped slot.

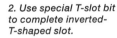
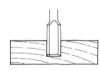
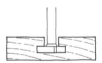

Slot will hold (left to right): T-bolts, T-slot nuts or carriage bolts.

MOVABLE CABINETS FOR SHOP STORAGE

by Joseph Beals

During the 10 years I worked in a cellar shop, I installed cabinets, drawers and open shelving wherever room allowed. The results were typical: I knew where to find everything, but there was little order to the method, and junk and dust were a chronic problem.

These movable cabinets keep tools stored, on slide-out shelves or in drawers, neat and dust-free. Casters make the base cabinets mobile while a cleat-mounting system allows the wall cabinets to be easily rearranged.

When I moved to a converted garage building, I left those built-ins behind. I packed tools, hardware and supplies into dozens of 5-gal. buckets, and I worked out of them for the next year until the new shop was at last functional. To avoid recreating the past, I designed a new storage system that remedies many of the usual irritations. I resolved to minimize any sort of generic storage that invites accretions of dust and junk. This meant little or no open shelving, no big drawers under the bench and no casual boxes or bins.

Finally, with the agony of moving so close behind me, I wanted a fully portable storage system. And I wanted a system that could be moved around easily.

Mobile cabinets

With these goals in mind, I built a set of floor and wall cabinets, as shown in the photo above, which offer exceptional utility in concert with a pleasing, traditional appearance; I also built special wall storage racks, as discussed in the sidebar on p. 28. The floor cabinets are mounted on casters and incorporate a series of guide rails for shelves or drawers. The wall cabinets hang from simple wall-mounted cleats (see the photo on p. 29) and include integral dadoes to allow any combination of plain or purpose-built shelving. To cut costs, I built the cabinets from a variety of wood species

Wall racks or clamps, lumber or shelves

With tools and hardware stowed out of sight in the new cabinetry, I was still left with a pile of clamps along one wall and a stack of lumber on the floor. My solution to both these problems was the same, as shown in the photo below and the drawing at right. The lumber rack is of identical construction to the pipe-clamp rack but built with more substantial members.

The racks are easily built by sandwiching spacer blocks between two vertical pieces to create mortises that hold the support arms. The support arms are angled on their lower edges, and the wedges that hold the arms in place have a matching angle on their upper edges. To secure the support arm, slip it into the mortise, push the wedge into the mortise below the arm and tap the wedge into place with a hammer. The arms can hold clamps, lumber or even shelves for storing other small items. —J.B.

Wall racks for clamps and lumber storage are easily made by sandwiching spacer blocks between a pair of vertical supports. Support arms slide into mortises and are secured with a wedge.

Wall racks _____

Simply constructed, this rack can hold lumber, pipe clamps or even shelves. Use 2x stock for the lumber rack; smaller stock is sufficient for other purposes.

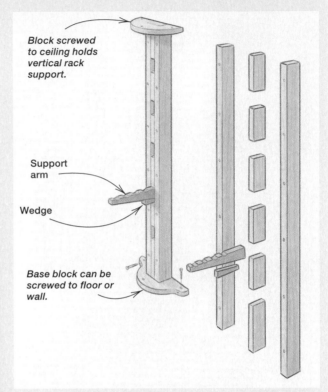

Block screwed to ceiling holds vertical rack support.

Support arm

Wedge

Base block can be screwed to floor or wall.

SUPPORT ARM DETAIL

To make detents that prevent pipes from rolling off support arms, drill 1-in.-dia. holes after clamping the support arms together, top edge to top edge, separated by a ¹/₂-in.-wide spacer.

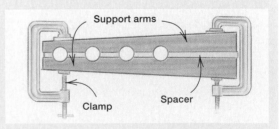

Support arms

Clamp

Spacer

Cleats allow cabinets to be moved—Strips ripped at 45°, one screwed to the wall and one to the cabinet back, make it easy to rearrange cabinets. A kicker screwed to the cabinet back near the bottom makes the cabinet hang plumb.

using leftover stock and cutoffs, including quartersawn white oak, black walnut, mahogany, elm and cherry. All cabinets include paneled doors, 1/2-in.-thick birch plywood backs and straightforward joinery.

Building considerations

Many woodworkers rely upon detailed, measured drawings as the final design stage, but that's like committing a melody to manuscript without opening the piano for a trial run. Unless you have an extraordinary ability to assess light and shadow, mass, proportion and function on paper, you risk building a sterile, techno-cratic piece. Remember that a final measured drawing merely records the component dimensions of a functional, aesthetically pleasing prototype.

I used molded frames, raised panels and polished finish to create a display for clients visiting my shop, but there are many simpler options. Pine frames made on the tablesaw, router table or entirely by hand, together with 1/4-in.-thick plywood panels and a paint finish are attractive and require no special tooling. A solid, flat panel, rabbeted around the perimeter to fit the frame grooves, is fully traditional and easy to make. If you are new to frame-and-panel work, these alternatives are a practical and satisfying introduction.

For a contemporary appearance, substitute plywood for frame-and-panel construction. Plywood cabinet sides can be grooved to house shelving or drawers, eliminating the guide rails required by a frame-and-panel carcase. Plywood has some drawbacks, how-

ever. A-C fir plywood is generally too crude for cabinetry, but 3/4-in.-thick birch plywood, which is the least expensive alternative, will cost about $45 per sheet and is best suited for a paint finish. Also, exposed plywood edges must be banded for a good appearance, even under paint. Commercial banding veneers with a hot-melt adhesive are easy to apply, but shopmade solid edge-banding is more robust, and it looks better.

Base cabinets

I built all the cabinets in multiples for maximum benefit of bench and machine time, but I'll describe the construction as if I were making only one of each, starting with a floor cabinet. I began with the frame-and-panel sides (see the drawing on p. 30). The stiles are equal in length to the height of the frame, but stile and rail widths and the length of the rails, are determined according to personal preference and the method of joinery. Mortise-and-tenon joinery, for example, requires additional length on the rails for the tenons.

I used a matched set of cope-and-pattern cutters on the shaper to machine the frame, but there are many other equally suitable methods (see *FWW* #86, pp. 76-79). To ensure accuracy, I took panel dimensions off a dry-assembled frame. After preparing the stock, I wasted the bulk of the bevel on the tablesaw, then finished fielding the panel with a panel-raising cutter on the shaper. (For more on machining panels, see *FWW* #94, pp. 65-69.)

When the two sides were assembled and cleaned up, I used my shaper to cut a rabbet

Base storage cabinet

Mobile storage cabinets make it easy to rearrange your shop layout. Uniform spacing of drawer guides makes all drawers and shelves interchangeable.

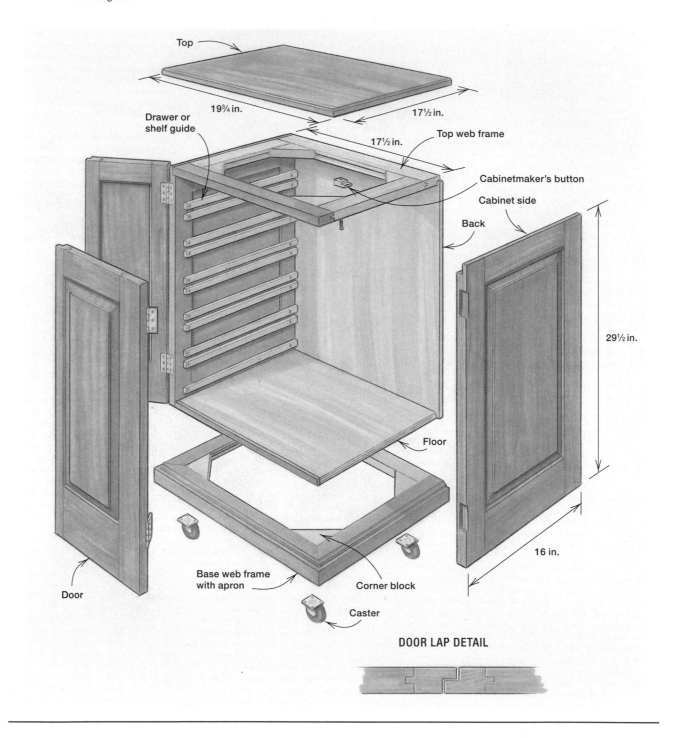

Top

19¾ in.

17½ in.

Drawer or shelf guide

17½ in.

Top web frame

Cabinetmaker's button

Cabinet side

Back

29½ in.

Floor

Door

Base web frame with apron

Corner block

16 in.

Caster

DOOR LAP DETAIL

Drawers and shelving are interchangeable because the guide spacing is uniform. Salvaged stock was used for the veneered drawer fronts.

Installing drawer guides is done with production-line speed by drilling counterbored screw holes and positioning the guides with spacers.

on the inside back edge to receive the plywood back and another across the inside top edge to house the upper web frame. Finally, I installed the maple drawer and shelf guides. The guide spacing is uniform, so any drawer will fit any space. And drawers and shelves are interchangeable, as shown in the photo above. To make this job accurate and quick, the guides are prepared in advance with counterbored screw holes and positioned with a series of spacers, as shown in the photo at right. That ensures consistent, square placement. I load the screws into their holes, and run them in with production-line speed.

An upper web frame keeps the top of the carcase square, provides fastening for the solid top and an upper stop for the doors. Notice that the front member of the web is full length and is the only part that need be primary wood. For a run of several cabinets, using secondary wood for the sides, back and corner blocks can save an appreciable amount of stock. A lower web frame is the load-bearing part of the cabinet base. The lower web is exposed on the front and both sides with the two front joints mitered for a better appearance.

Both web frames are of traditional construction, as shown in the drawing on the facing page. I used a shaper to machine the full-length grooves, and I cut the tongues on the tablesaw. I also splined the lower web miter joints for strength and convenience of

positioning during assembly. To avoid juggling eight parts at once, the web frames are first glued without corner blocks. When the glue has set, the corner blocks are installed in a second operation.

A skirt on the front and sides of the lower web frame gives visual mass to the cabinet base and shrouds the casters. After the skirt is applied, I used a pair of cutters on the shaper to machine a simple reverse ogee (cyma reversa) molding detail on the front and sides, giving a graceful transition from the carcase to the base. Finally, a plywood shelf is affixed to the top of the lower web frame, with the front edge-banded with the cabinet's primary wood. The shelf provides a cabinet floor and positive positioning for the

Wall storage cabinet

These wall cabinets are also mobile. The mounting cleat system on the back is solid yet can be disassembled quickly and easily.

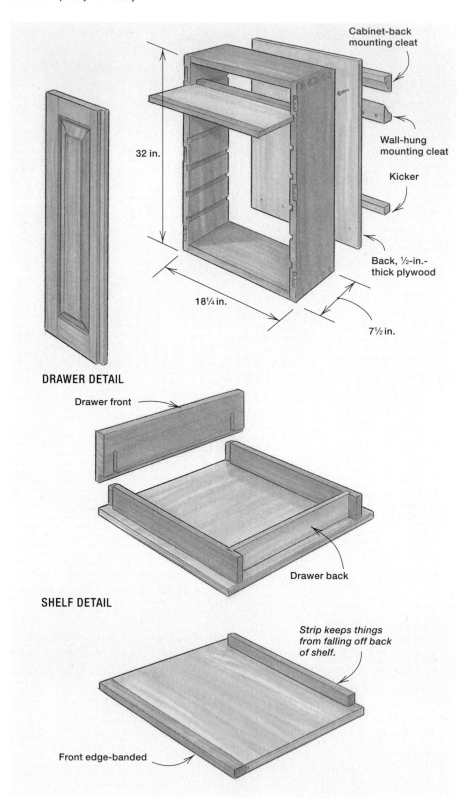

Cabinet-back mounting cleat

Wall-hung mounting cleat

Kicker

Back, ½-in.-thick plywood

32 in.

18¼ in.

7½ in.

DRAWER DETAIL

Drawer front

Drawer back

SHELF DETAIL

Strip keeps things from falling off back of shelf.

Front edge-banded

cabinet sides, and the front edge serves as a lower door stop.

All the cabinet components are screwed together, so assembling the cabinet is quick and easy. The two sides are fastened to the lower web frame with screws driven up from below, just inside the skirt. The upper web frame is screwed down into the sides from the top and the plywood back is screwed into the rabbets that house it. I used no glue in the assembly, which makes it possible to take the carcase apart for any reason. I used shop-grade elm for a serviceable top on all floor cabinets. The tops are given a half-round profile on the front and sides, and they're fastened to the upper web frame with traditional buttons (see the drawing on p. 30).

The cabinet shelves can be solid stock or sheet goods, as preference dictates. I used $1/2$-in.-thick birch plywood, banded on the front to match the cabinet wood. The shelves pull out easily on the guides, and thin cleats glued to the back keep things from falling off the back edge.

To keep the design simple, I built the drawers as a box fastened to a shelf, as shown in the detail on the facing page. The two sides engage the front with sliding dovetail joints, and the front of the shelf fits in a rabbet on the drawer front. I screwed the sides and back to the bottom from below. The cabinet will hold six shallow drawers, but deeper drawers can be made by doubling or tripling the spacing module.

Building wall cabinets

Construction of the wall cabinets is simplicity itself (see the drawing on the facing page). The solid carcase can be assembled in a variety of ways (see *FWW* #104, p. 75), but through-dovetails offer the strongest and best-looking joint. I cut all dovetails by hand, which took less than an hour for each cabinet. Before the carcase was assembled, I used a dado set on the radial-arm saw to cut six shelf dadoes in each side. Using steel shelf standards or a series of holes for shelf

support pins would provide a greater range of spacing options and a cleaner look, but my prior experience with these methods was unsatisfactory. A fully housed shelf never tips out nor does it require a store of mounting pins or brackets, which are typically missing at the moment of need.

Finally, I machined a rabbet on the inside back edges to house the plywood back. The rabbet is full length on top and bottom and is stopped a $1/4$ in. short of the ends on each side. I cut the rabbets on the shaper and cleaned the stopped ends with a chisel after the carcase was assembled. The back rabbet could also be routed with a bearing-guided bit or on a tablesaw. When the back is in place, a hanging cleat is fastened near the top (see the photo on p. 29), and a kicker is fitted near the bottom to keep the hung cabinet vertical.

Frame-and-panel doors

All cabinets are provided with a pair of narrow, paneled doors. A single-wide door might seem simpler, but the sweep can be awkward, especially on a floor cabinet in a restricted space. The doors are constructed like the floor cabinet sides and for appearance, have the same dimensional proportions of stiles and rails. Notice, however, that the two inside stiles are half width and give the appearance of a single, full-width stile when the doors are closed. The stiles are also half-lapped, as shown in the door lap detail on p. 30, such that a single catch on the right-hand door keeps both doors closed. I cut the inside stiles wide to assemble the doors and machined the lap joint after assembly.

I used aniline dye to stain all but the cherry cabinet. Because cherry darkens so rapidly and dramatically, it is generally better not to color the wood under the finish. All cabinets received a half dozen coats of shellac, and after the last coat is well rubbed out, I applied a beeswax polish for a soft, lustrous finish.

VACUUM HOLD-DOWN/ ROUTER TABLE

by Mike M. McCallum

Vacuum side has plenty of clamping power—The vacuum surface (left) clamps a drawer front as the author sands its face. An open hole on the edge of the table (right) shows where he connects his shop vacuum when he's routing.

When I'm constructing a set of custom cabinets, I frequently need an extra pair of hands, especially when I'm sanding drawer fronts or drilling odd-shaped pieces. Occasionally, I also need a table-mounted router. More often than not, I require that router table or those pair of hands at a job site. After putting up with cobbled scraps, make-shift clamps and excessive router dust one too many times, I came up with a design for a router table that's also a vacuum hold-down. Using scrap materials, I built the table so that I could easily disassemble it for storage or transport.

I call my knockdown platform a super router/hold-down table for a couple of reasons. First, it's stout, turning my router into a light-capacity shaper. Second, it enables my shop vacuum to serve dual functions by providing suction for the hold-down surface or collecting dust from the router table. And while I don't use the hold-down to freehand-rout large workpieces, I do rely on its substantial holding power for most of my sanding and finishing work (see the photo at left).

Design and construction

The dimensions of the hold-down table are not critical, but be sure to adjust for the size of your work area, vacuum hose and router. I made my table out of $^5/_8$-in. high-density particleboard and covered exposed surfaces with scraps of plastic laminate. The top is removable, so I can use the vacuum table on my benchtop. I stiffened the table's top and bottom by gluing on a particleboard framework, as shown in the drawing on the facing page. The top and bottom frames hold the sides and center divider in place without fasteners, allowing easy knockdown of the unit. After assembling the top and bottom oversized, I trimmed the parts square. I laminated all the pieces, and then I bored two holes in the edge of the table, so I can connect my shop vacuum to either the router-table or hold-down side (see the photo at left). To power the table, I ran a heavy-duty extension cord to a 4x4 electrical box and mounted the box's stud bracket to the inside of the platform. The box houses switched receptacles for both the router (or sander) and shop vacuum. I also added a plywood shelf (see the photo on p. 36) to the table to hold tools, bits, guide bushings and adapters. I ordered most of these accessories through MLCS Ltd. (PO Box 4053 C-13, Rydal, Pa. 19046).

The router-table side

I made a clear window for the router table from Lexan, which I recycled from a computer-store display. The window is a good safety feature because it lets in enough light to see the collet when I'm adjusting the bit height or using the router table. The router insert is a standard one—it fits whatever bit I'm using. Oak Park Enterprises, Ltd. (Box 13, Station A, Winipeg, Manitoba, Canada R3K-129) carries complete bearing and insert kits for various router models. To prevent vibration and flush the router-table surface, I oiled the insert's bearing surface and then caulked the face of the insert with silicone before I placed it into the flange that I had routed into the tabletop. I clamped the

Vacuum hold-down/router table

All ⅝-in. high-density particleboard, unless noted

Hold-down holes, 1/64 in. dia., have chamfered edges. Pattern matches tree-shaped chamber.

Laminate top

Router insert

Rout tree-shaped chamber at stepped depths (see air channel detail).

Particleboard framework

Plywood shelf

Vacuum hook-up holes

Caulk Lexan window into ¼-in.-deep routed flange.

Sides, 13h x 14w

Heavy-duty extension cord connects to 4 in. x 4 in. electrical box for switched receptacles.

Pine rail, 1 in. x 5 in.

Top and bottom frames are 1½ in. x 17 in. x 30 in. Overall height of table is 16 in.

Pine shelf cleats

Apply plastic laminate to exposed surfaces.

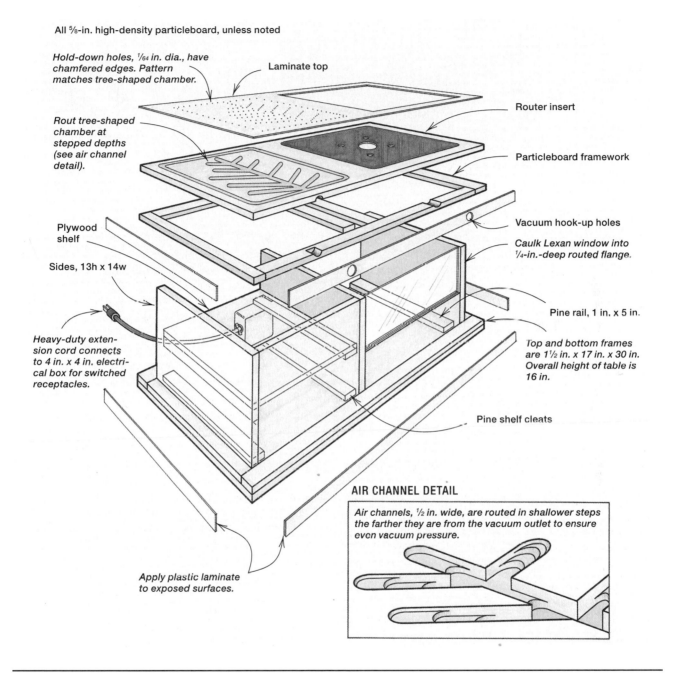

AIR CHANNEL DETAIL

Air channels, ½ in. wide, are routed in shallower steps the farther they are from the vacuum outlet to ensure even vacuum pressure.

To wire his table, the author fed a 4-in. work box for a pair of switched receptacles: one for the vacuum, one for a router or sander. A shelf at the back of the table holds tools and accessories. He drew outlines of the items and routed recesses to hold each shape, which reminds him when something's missing.

assembly flat on my tablesaw. In use, I estimate that my 5-gal. collector sucks up about 50% of the dust from the router. I'm sure that with a few provisions, such as adding intake holes, I could improve the dust-collection capability considerably.

The hold-down side

On a piece of tracing paper, I drew an evenly spaced tree-shaped hole pattern that was suitable for my hold-down needs. After I transferred the tree pattern onto the particleboard top, I freehand routed the air channels for the vacuum chamber. To ensure an even vacuum across the hold-down surface, I routed the channels at ascending depths (see the air channel detail in the drawing on p. 35). I based the stepped depths on vacuum-drop ratios for the chamber volume. If you're using plywood for the table, paint or seal the routed pattern to prevent air leaks before you glue on the laminate. To make the hole pattern in the laminate, I

first placed a clear piece of plastic over the routed chamber and poked out a hole pattern to follow the air-channel shape. Then I used a crayon to rub the hole locations onto the laminate. When boring through the laminate, use a tiny bit (I used a $\frac{1}{64}$-in. twist drill). The small orifices, through the Venturi principle, increase the vacuum. Finally, lightly countersink the holes.

Using the vacuum hold-down

As long as my workpiece has a flat surface to put down on the hold-down table, I've found that there's plenty of suction— enough to grip a piece of low-grade plywood. To increase the holding pressure, you could also block off holes that are not covered by the workpiece. On rough surfaces, I take a $\frac{1}{8}$-in.-thick piece of closed-cell plastic (shipper's foam) to make a gasket. With a utility knife, I cut out an appropriate shape that still allows the vacuum to suck the workpiece down. I use a couple of pieces of masking tape to hold the gasket to the table. If you need to hold down a sphere, an odd shape or a piece with a very uneven surface, you can make a holder as follows: Sculpt out a Styrofoam gasket for the shape you want to secure. In a well-ventilated area or outdoors, heat up a piece of nichrome or small stainless-steel wire with a propane torch, so you can make a series of holes through the gasket. Tape the Styrofoam to the hold-down table, and you've got a fairly quick clamp to hold just about any shape that you have to sand or drill holes in. So far, I've been delighted with the possibilities of the hold-down table. In fact, I'm working on a sliding saw table that uses a similar vacuum hold-down system.

SHOOTING BOARD AIMS FOR ACCURACY

by Ed Speas

Perfect miters— Guided by Ed Speas' shooting board, a Lie-Nielsen #9 miter-plane easily shaves a 45° miter on molding. The fence is reversible, so the fixture can handle left- and right-hand cuts.

Fitting miters has been every woodworker's problem at one time or another. Whether you are making a picture frame or joining molding, if your angle of cut or your piece lengths are not perfect you have to repeatedly shave a smidgen to get a tight joint. Although a chopsaw or a tablesaw can save time and effort, it may not be the best choice for extremely clean and accurate cuts. If you use a handsaw it tends to wander if not precisely guided. And even then, I don't know too many folks who can really get consistent forty-fives with a hand miter-box alone. Trimming 90° cuts can also be a problem. A saw blade, hand or power, rarely leaves a smooth enough surface. If you sand the endgrain, again you risk introducing error.

You can eliminate these difficulties by using a simple fixture called a shooting board. When guided by a shooting board, a plane with a razor-sharp, lighly set iron can accurately slice off wispy thin shavings, as shown in the photo above. And the end-grain will be left with the smoothest surface possible. To use one of these fixtures, you first place a workpiece against the fence and lay a handplane on its side with the sole against the edge of the base. Butt the work up to the plane sole, and then push the plane by the work in several passes.

The shooting board I use is an adaptation of an old bench hook, or sawing board. I made this combination bench hook/shooting board to either hold stock while sawing (see the photo on p. 38) or to precisely plane

Fixture doubles as a bench hook—To convert the shooting board to a bench hook for 90° sawing, the author simply removed the miter fence (here resting in the bench trough).

the ends of stock. One of the fixture's unique features is its removable 45° fence, which makes it both a miter and a right-angle shooting board. The fence is reversible as well, so I can pare miters from the left or the right side, a great advantage when I need to work each half of a joint in molded work.

Making the fixture

My shooting board consists of a rectangular base and fence, a triangular miter fence and a hook strip, which serves as a bench stop and a clamping cleat. I made all of the parts out of medium-density fiberboard (MDF). To get the 1-in. thickness I wanted I first laminated two pieces of $1/2$-in. MDF, about 9 in. by 25 in. Next I cut out pieces in the dimensions shown in the drawing on the facing page, making sure all the corners were exactly square and the 45° angles were dead accurate, not just close.

When assembling the shooting board, I was concerned about how much pounding the fixed fence would take. That's why I both glued and screwed it to the base. I attached the hook the same way. First, I pre-drilled and countersunk the screw holes. Next I aligned each piece with a square and glued and clamped it to the base. Then I fastened each in place with bugle-head drywall screws.

The removable miter fence registers against the fixed fence and is held down by a snug-fitting pin. I used a $1/4$-in. bolt with the head cut off for the pin. As an alternate, a hardwood dowel would work, though I suspect over time the pin would become loose. Because the location of the pin and the size of its holes are critical, I bored the holes with my drill press. First I drilled a $1/4$-in. pin hole clear through the miter fence in the location shown in the drawing on the facing page. Next I clamped the fence to the base in its right-hand position so I could drill through the pin hole into the base. I flipped the miter fence and did the same thing to make the hole for the left-hand position.

Using cyanoacrylate (Super) glue, I secured the pin in its hole, letting it hang out about $1/2$ in. on the underside of the fence. For aesthetic purposes I plugged the top $1/4$ in. of the fence hole with a dowel. I rounded over the end of the pin (bolt) and then tried its fit in the base hole. With the shooting board together, I clamped it in my bench vise. Then I laid my plane on its side and took a shaving off the shooting edge, both sides. Because a plane iron does not go all the way across the sole, the iron leaves a slight recess (rabbet) along the base. This is necessary for proper registration of the plane. I had to shim the board $1/8$ in. so the plane would shave the full edge. After dusting the fixture off, I finished the whole thing with oil. Later I waxed the shooting board to keep it slick and clean.

Shooting square cuts and miters

To use the shooting board, clamp its hook in an end vise to keep the fixture stable. Make sure your bench is dead flat or lay down a flat auxiliary table before clamping the fixture. While steadying the workpiece, hold the plane with a firm grip and keep it tight against the edge of the shooting board as you take multiple passes. Use the largest bench plane you have. A Stanley No. 07 or 08 jointer plane works best, but a No. 05 jack plane will also do, as long as it has a sharp iron and its sole is true and square to the plane's body. Even better, you can use a

Shooting board assembly

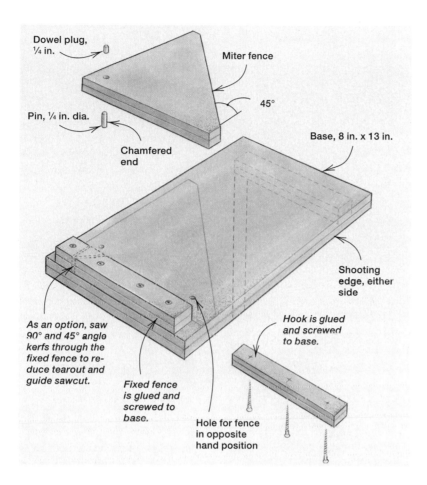

Dowel plug, ¼ in.

Miter fence

45°

Pin, ¼ in. dia.

Chamfered end

Base, 8 in. x 13 in.

Shooting edge, either side

As an option, saw 90° and 45° angle kerfs through the fixed fence to reduce tearout and guide sawcut.

Fixed fence is glued and screwed to base.

Hole for fence in opposite hand position

Hook is glued and screwed to base.

miter plane, which resembles an oversize block plane and is specifically meant for shooting (see the photo on p. 37).

When shooting the end grain of a right angle cut, it's a good idea to first knife an edge line right around the board, which will prevent tearout. Then plane to the line. When shooting 45° angles, tearout is rarely a problem. In this mitering mode the shooting board can trim very small amounts. This is crucial when fitting a lipping around a veneered panel, for example, where the length of the lipping from inside miter to inside miter has to be exactly the length of the side of the panel. Because the fence pin serves as a pivot point, you can also adjust the angle of cut slightly to bisect a corner that's not quite square. Just insert a paper shim where needed between the fences. I have a stack of old business cards that work great for this.

THREE

Jigs for Routers

Like a car without a road or a computer without programming, the router isn't much without a jig. Together, however, they make a mighty potent pair.

The router is such a small and deviously simple tool that the way it has transformed the modern woodshop boggles the imagination. It's not much more than an electric motor with a rotating shaft that accepts cutting bits. But its ease of use and versatility make it the tool of choice for an extensive list of common shop jobs. It will make any molding. It will mill any shape from profiles to compound curves. It will cut mortises, tenons, biscuit slots, and other joints. You can even carve with it and use it to shape on a lathe. It can be adapted to almost any cutting role—but not by itself. Only through jigs can the router realize its potential. Without jigs, it's a mere molding cutter.

A router jig is a guidance system. It controls the path of the cut, guiding the router past the workpiece or the workpiece past the router. It can be as simple as a fence with a stop and as complex as a milling machine with a few hundred parts. It all depends on the complexity of the cutting you aim to do.

This chapter can only begin to offer some ideas for guiding routers with jigs. Three articles on router tables show the range of possible setups for this most basic router jig. Tables turn the router into a small shaper, which offers its own endless list of cuts, from raised panels to sliding dovetails and tenon shoulders. Then several articles show a range of ways to cut joinery, from mortises and tenons to dovetails and dadoes, and all of them in one "joint-making machine." Finally, an article on a very specialized jig for flush-trimming edgebands completes the section. Unlike its more involved brethren, this jig does only one thing, but it does it so quickly and cleanly, you won't regret making it.

These jigs are only the beginning of the router's potential, examples of how woodworkers have uniquely adapted their routers to specific tasks. There is almost nothing the router can't cut without a good jig to guide it. It is truly a power tool limited only by the extent of your imagination.

NO-FRILLS ROUTER TABLE

by Gary Rogowski

Remember the commercial about the knife that sliced, diced and performed a myriad of other tasks, even gliding through a tomato after cutting a metal pipe? Well, that's what a router table is like. You can cut stopped and through grooves, dadoes, rabbets and dovetailed slots. You can raise panels, cut sliding dovetails, tenons and mortises. It's no wonder that many woodworkers can't imagine working wood without one.

But router tables can be expensive. In one woodworking catalog, I saw a number of packages selling for between $250 and $300. I'd rather spend my money on wood. That same money would buy some really spectacular fiddleback Oregon walnut.

I've been building furniture for years, and my bare-bones router table has given me excellent, accurate results. The router table in the photo at right is a variation that is inexpensive, simple to construct and extremely versatile. It's a simple, three-sided box made from a half-sheet of ³/₄-in.-thick melamine with the front left open for easy access to the router. I made mine with a top that's 24 in. deep by 32 in. wide, which keeps it light enough to move around yet big enough to handle about anything I'd use a router table for. It's 16 in. high, which is a good height for placing it on boards on sawhorses or on a low assembly bench.

Biscuits and dadoes join parts

When you buy the melamine, make sure the sheet is flat. And buy it in a color other than blinding white—it's tough on the eyes.

The melamine I used had a particleboard core. Biscuits are stronger than screws in particleboard, so I joined the two sides to the top with #20 biscuits. To make the cuts in the underside of the top, I took a spacer block 5 in. wide, aligned it with the end of the top and set my plate joiner against it for the cuts. The width of the block determined the overhang of the top. Marks on the spacer block gave me my centers.

The biscuit joints probably would have been plenty strong by themselves, but I

wanted to add a little extra strength to the joint. So I decided to dado the underside of the top for the sides. I couldn't dado very deeply, though, or the biscuits would have bottomed out. I settled on a $^1/_{16}$-in.-deep pass centered over the biscuit slots (see the photo at right). Before cutting the dado, however, I dry-fitted the sides and top with biscuits in place to check alignment. Then I scored heavily around the edges of the side pieces with a marking knife and routed the shallow dadoes.

Before gluing the sides to the top, I rabbeted the back edge of the two sides for a $^1/_4$-in. panel to strengthen the table and prevent it from racking. Then I glued the

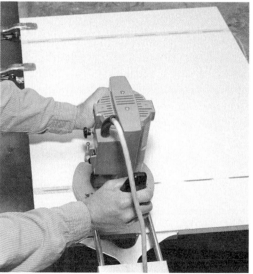

Shallow dado increases glue surface. To strengthen the joints between the sides and top, the author routs a dado $^1/_{16}$ in. deep in the underside of the top directly over biscuit slots.

Fiberboard back prevents racking. Although it's only ¼ in. thick, the fiberboard back greatly strengthens the table. The fiberboard is glued and screwed into a rabbet all around the back of the table.

sides to the top one at a time, using battens to distribute the clamping pressure. I made sure each side was square to the top and waited for the glue to set up.

I used a router and rabbeting bit to cut a stopped rabbet in the back edge of the top. Then I glued and screwed down the ¼-in. medium-density fiberboard (MDF) back panel (see the top photo at left). Hardboard or plywood would have worked as well.

I use a fixed-base router in my router table because it's lighter than most plunge routers and won't cause the table to sag over time. Also, it's much easier to change bits. I just drop the router motor out of the base, change bits, reinstall the router and I'm back to work.

I attached the router base to the underside of the tabletop with machine screws that go down through the top into the tapped holes in the router base. To mark the location of the screw holes, I removed the router sub-base and made pencil marks on the top. Then I drilled and countersunk holes into the tabletop.

With the base attached to the table, I marked out where the bit hole should go and drilled a ¾-in. hole into the table. I put a 2⅛-in.-dia. chamfer bit in my router— the largest bit I have. I started the tool and gradually moved the bit up and through the tabletop (see the bottom photo at left).

To prevent workpieces from diving into this hole when using small bits, I made a set of inserts that fit in a shallow recess around the bit hole. Holes in the inserts accommo-

Cut the hole with a router bit. With the router base screwed to the underside of the top, the author advances his largest bit through the table. Go slowly.

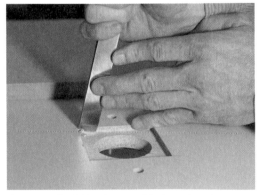

A recess for interchangeable inserts—A plunge router and chisel make short work of a recess in the tabletop that accepts inserts for different-sized bits.

date bits of different sizes with minimal clearance. I routed out the rabbeted recesses for the inserts first, using a plunge router guided by a straightedge. I squared the corners with a chisel.

I made the inserts of ¼-in. tempered hardboard. Their square shape keeps them from spinning during use and makes them easy to fit. I cut a bunch of them on the tablesaw and then sanded each to a perfect fit on a belt sander.

L-shaped fence provides dust collection

The fence I've always used might be called low tech, but there's really no tech to it at all. It's simply a straight, wide, flat piece of wood jointed so that one edge is square to a face. I clamp it to the router table wherever I need it. The fence doesn't have to be parallel to a table edge to work. When a bit needs to be partially hidden for a cut, I use another board with a recess cut into its face.

The only thing my primitive fence lacks is dust collection. Hooking up a vacuum or a dust collector just won't work in some situations, such as when I plow a groove. But with other operations—raising a panel, rabbeting a drawer or box bottom, or cutting an edge profile—having a fence with a dust port can really help clear the air.

The fence I built for this router table is made of two pieces of ³⁄₄-in.-thick MDF about 4½ in. wide and 49 in. long rabbeted together to form an L-shape (see the top photo at right). I cut a semi-circular hole at the center of each for dust collection. This allows for better pickup. I also routed slots in the vertical part of the fence so I could attach auxiliary fences for specific operations, such as raising panels or rabbeting. Once these slots are routed, the two pieces of the fence can be glued together. Make sure the fence clamps up square because virtually everything you use the table for depends on it.

To create sidewalls for the dust-collection hook-up, I added two triangular-shaped pieces of ³⁄₄-in.-thick MDF to frame the dust-collection port (see the bottom photo at right). I glued these triangles in place on either side of the dust holes, just rubbing them in place and letting them set up without clamping. After the glue had cured, I

Clamp the fence square. Adjust the clamps to get the two pieces square over the entire length of the fence.

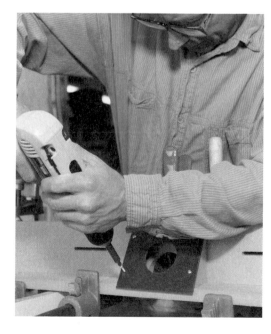

Screw the dust-collection port to fence. Smear a bead of glue along the two triangular sidewalls. Drill holes and screw the hardboard back to them.

filed the triangles flush with the fence, top and bottom.

To complete the dust-collection hook-up, I measured the diameter of the nozzle on my shop vacuum and cut a hole to accommodate it in a piece of $1/4$-in. hardboard. I left the hardboard oversized, clamped it to the drill-press table and used a circle cutter on my drill press. Then I cut the hardboard to size and glued and screwed it to the two triangular walls.

Using a closed auxiliary fence—Routing away from the fence calls for auxiliary pieces butted tightly together to form a smooth, continuous surface.

Auxiliary fences solve specific problems

A two-piece auxiliary fence can be used to close up the area around the bit when routing profiles, rabbeting or performing similar operations. This way, there's no chance of a small piece diving into the gap between bit and fence. And with a smaller opening around the bit, the dust collector or vacuum will work more efficiently. When the fence is situated back from the bit, such as when mortising, another set of auxiliary pieces can be used, so there's no gap between the two halves (see the photo at left).

I made the auxiliary fence from two more pieces of MDF. The auxiliary fence is drilled and countersunk for machine screws that ride in slots cut in the main fence. I use nuts and washers to tighten the two pieces in position.

When using the auxiliary fence, I close the two halves around the moving bit to provide a custom fence. When I'm done with it, I can set the fence aside for future use or just cut it off square and use it again. Closing the fence into a bit with a diameter that's less than the thickness of the fence will not open up the back of the fence to the dust-collection port. In this situation, I pivot the fence through the spinning bit before setting the fence for depth of cut.

Make sure that the outfeed side of the fence doesn't stick out any farther than the infeed side. If it does, it will prevent you from feeding your work smoothly past the bit. If your work catches on the outfeed side of the fence, easing its leading edge with a file or a chisel may help. If it doesn't, you can always shim the infeed side with slips of paper.

Another router table problem I've found is what to do with large upright pieces, such as panels cut with a vertical panel-raising bit. The solution is to screw a taller auxiliary fence to the main fence. The fence can be pivoted right into the bit, so there's no gap on either the infeed or outfeed side of the bit, yet there's dust collection behind the bit.

CAST METAL ROUTER TABLES

by Mark Duginske

Turning a router upside down increases its versatility dramatically. A router table provides both a large perpendicular reference surface (the table) and a long horizontal reference surface (the fence), making it possible to perform many tasks much more safely and with greater ease than is possible with a hand-held router and a clamped, makeshift fence or a bearing-guided bit. The router table has changed woodworking so much in the past decade, in fact, that a host of cottage industries have sprung up to support the demand. Once it was a somewhat obscure denizen of professional shops that couldn't justify the expense of a shaper, but the router table is probably rivaled in popularity today only by the tablesaw.

Many woodworkers made a first router table out of standard $3/4$-in. plywood, screwing the router directly to the plywood and then just running the bit up through the plywood. A refinement to this setup was the addition of a lift-out, plastic insert, which made changing bits considerably easier. Then, as router tables started to become available commercially, the plastic inserts were followed by leveling screws for the inserts, fancy fence setups, dust control hookups and elaborate base cabinets. Still, most of these tables have been and continue to be made of plywood or medium-density fiberboard (MDF), usually with a skin of plastic laminate. None of these plywood or MDF router tables have incorporated an adequate (non-wearing) miter-gauge slot in the table, and very few have miter-gauge slots at all.

Porta-Nails' optional pin-router attachment ($50) makes it possible to template rout any number of identical irregularly shaped pieces. The attachment is simple to set up and to adjust.

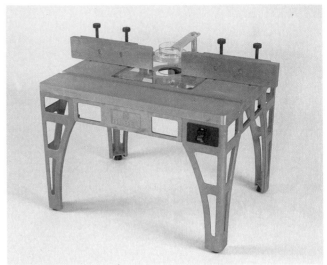

From left to right: Porter-Cable Model 695, Woodstock International W2000 Rebel, NuCraft Tools cast-iron router system and Porta-Nails universal router table. Each of these tables has a cast-metal tabletop and a miter-gauge slot.

In the past year or two, a number of metal router tables have been manufactured that attempt to address this and other inadequacies of conventional plywood and MDF router tables. Limiting the field to cast (as opposed to stamped metal) tables that take a conventional miter gauge ($^3/_8$-in. by $^3/_4$-in. bar), I identified four models (see the photos above and on the facing page). I spent a few weeks working with all of the tables last December, using them for a variety of applications. This is what I found.

Porter-Cable Model 695

Porter-Cable's benchtop router table was the cheapest of the four. Though it lists for $390 including the fence, I was able to pick it up for about $135 at my local hardware store. It took only a few minutes to assemble. Stamped-steel legs screw directly to the underside of the table, placing the tabletop 10 in. off of your work surface (see the bottom photo on the facing page). The 16-in. by 20-in. table is sufficient for small- to medium-scale woodworking.

The die-cast aluminum table is the only one of the four that does not use a table insert. This could be a liability if you change bits frequently because it's awkward to reach under the table to loosen the collet. I intend to keep the table and use it as a dedicated rabbeting station. Because all I'll have to do is change the height of the bit or reposition the fence, for my purposes, it's not a prob-

lem. Even so, Porter-Cable would do well to add a removable plastic insert to this table.

No miter gauge comes with the table, but my Delta miter gauge fit in the slot with only a few thousandths of an inch of play side to side, which is plenty precise for woodworking. A safety starting pin, which gives you a surface off of which to pivot the workpiece when you're working with a bearing-guided bit and no fence, comes standard. A safety guard for when the fence isn't in use also comes standard.

The standard fence, a stamped-metal, two-piece affair (like a shaper's), is this table's weakest link. The range of adjustment is so small—even with the fence back as far as it will go—that it restricts the operations you can perform. For example, when fully retracted, the front of the fence sits only about $^1/_{16}$ in. behind the center of the bit, which makes routing dadoes impossible. If I were to buy this router table as my main router table, I'd do what I've done in the past and just build a pivoting wooden fence with a round dust port near the cutter to connect my shop vacuum.

The table's greatest strength (besides its price) is its portability. For small-scale architectural woodworking, in situations where lugging around a bigger router table would be a hassle, this unit is ideal. It's light enough to move easily, and it's also small enough to store under a bench or even in a good-sized tool storage cabinet.

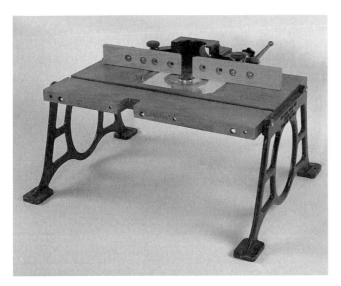

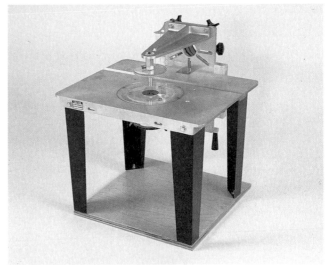

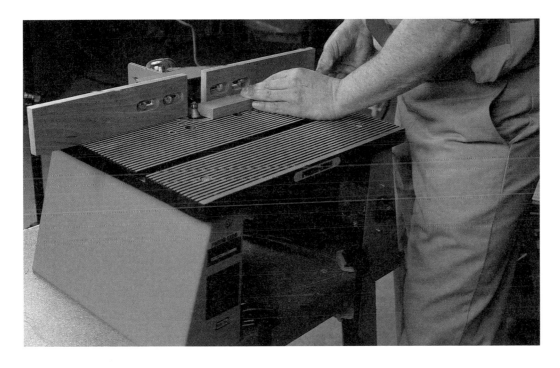

Simple, inexpensive and portable, the Porter-Cable Model 695 is a good buy for less than $150.

Woodstock International W2000 Rebel

With its cast aluminum legs and stretchers, the Rebel is a rigid, solidly built router table (see the photo on p. 50). It lists for about $230 (you can find it for around $200), and the inserts start at $19.95. An optional safety switch costs $12.95. The table surface is 18 in. by 24 in. and made of cast aluminum like the rest of the unit. Its 17 in. height was too tall for my workbench, but a separate stand is easy to make. Adjustable rubber feet screw into the bottom of the

aluminum legs, so it's easy to level the unit on an uneven surface. If I were to buy this table, I'd remove the rubber feet and screw the router table securely to a base cabinet.

A miter gauge, safety guard and a fence system consisting of two independent halves come standard with the table. Each half of the fence system is secured to the top of the table with bolts that slide in two slots. This style fence is convenient on a shaper but overkill on a router table, requiring more time and effort than it's worth. Apparently,

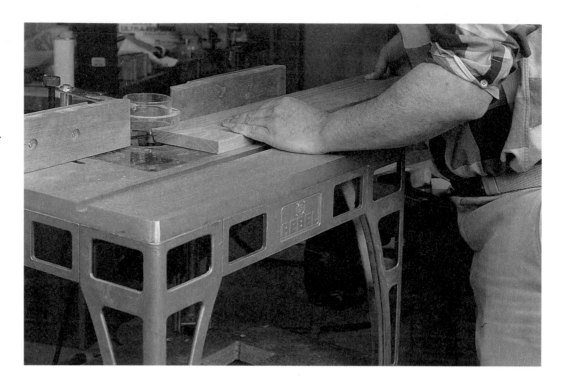

Cast aluminum tabletop, legs and stretchers give the Woodstock International W2000 Rebel a solid chassis. Because of its 17-in. height, it's best used on a dedicated stand.

from looking through the owner's manual (this manual was the best of the lot), this fence configuration is designed to accommodate two Incra Jigs, one on each side of the bit. Whether you use such aftermarket jigs will determine how important this feature is to you, but I found the whole fence setup inconvenient.

When I first tested it, this table's real liability was its plastic insert. With my Porter-Cable 690 mounted to it, the $1/4$-in.-thick insert sagged nearly $1/16$ in. at the middle of its 11-in.-sq. surface. The idea of hanging a $3^1/2$-hp router from this insert made me nervous.

Apparently, I wasn't the only one who noticed this problem. When I called Woodstock to mention it, the product manager told me that new inserts, one of $1/4$-in. aluminum and one of $3/8$-in. clear plastic, were already in the works and that they'd send one out to me as soon as they became available. A couple of weeks later, I received the aluminum insert. I mounted the same router to it and even pushed down near the center of the insert, but there was no deflection.

A nice feature of this aluminum insert is that there's a round $4^9/32$-in. hole in the middle that accepts snap-in, friction-fit plastic inserts. The rigid aluminum insert supports the router, but the plastic snap-in inserts make it possibe to have a bearing surface for the workpiece right up to the bit, resulting in safer and more precise routing (because it's easier to set bit height more accurately).

Woodstock's other new insert, is made of $3/8$-in. clear plastic and is rabbeted around its edges so that it sits flush with the aluminum table. It is not drilled out at all, so if you get this insert, you'll have to drill and countersink holes at the corners for the hold-down screws, and run your bit up through the plastic for a center hole.

NuCraft Tools cast-iron router system

From tabletop to extension wing to ornate legs to a shaper fence, all of NuCraft Tools' router-table components are made of premium-grade, fine-grained cast iron and plenty of it. Both the table and extension surfaces and the ribs beneath are at least $3/8$ in. thick, and there's more than enough iron surrounding the miter slot. Anyone fond of turn-of-the-century machines will appreciate NuCraft Tools' marriage of 19th- and 20th-century technologies.

Sixty-five pounds of cast iron goes a long way toward damping router-induced vibrations. NuCraft Tools Model 100 router table can either be attached to a standard contractor's or cabinet-style tablesaw as shown here, or it can be used as a stand-alone with cast-iron legs also sold by NuCraft.

The tabletop itself (Model 100; $295) is 18 in. by 27 in., so it attaches neatly to the standard Delta, Sears and Taiwanese tablesaws as an extension table (see the photo above). The tabletop also can be used as a stand-alone with cast-iron legs, which are sold separately (Model 200; $125). Also available is a 12-in. by 27-in. extension wing (Model 300; $155), which can be attached either to the left side of a tablesaw as an extension wing or mounted to the stand-alone router table along with the cast-iron legs.

This router table's greatest asset *and* its greatest liability (besides its price) is its weight. The tabletop weighs in at 65 lbs.; the extension wing adds over 45 lbs.; and the legs, at 15 lbs. each, bring the total to over 140 lbs. Even with Porter-Cable's biggest, variable-speed monster industrial router and a conventional panel-raising bit humming at however many thousand rpms, this table's not going to vibrate, which besides the miter-gauge slot, is one of the best reasons to consider a metal router table. By same token, if you have to move your router table around often, you're going to need a helper or a good back and a back-

support brace; it's no fun hauling this setup around. Attaching just the main table to my tablesaw was a two-person job.

The plastic insert plate is $^3/_8$ in. thick and has holes for the hold-down screws at the corners. This was the only table-insert combination without a leveling mechanism. Because the cast iron is so fine-grained, though, I was able to drill and tap holes near the corners for Allen screws.

The most recent NuCraft Tools accessory is a heavy-duty shaper-style fence (Model 400; $110). My prejudices aside (I don't think this style fence makes all that much sense for a router table), the fence was ground in such a way that the two halves of the fence both tilted in toward the bit. This caused the workpiece to catch on the lip of the outfeed fence. I pulled the workpiece away from the fence, finished the pass and then examined the workpiece. It was slightly convex and the rabbet was cut deeper on the portion that was routed before it reached the outfeed table. Curious whether I'd received a lemon, I obtained another fence. The results were the same—all the more reason to use a simple shop-built fence. If you really like this style fence, you could have the fence halves reground (at some expense), or you could joint their wooden faces. But for $110, I'd expect better.

Other accessories will be available soon, including a horizontal milling attachment.

Porta-Nails universal router table

The 16-in. by 20-in. cast-aluminum table by Porta-Nails sells for $359, which might seem a little high, but this table is the most versatile of the four I looked at. The table itself is nothing fancy—just a well-machined, heavy-duty, aluminum casting with a miter-gauge slot, leveling screws and a round $3/8$-in. plastic insert. The legs are stamped steel, and they're attached to a wooden base that is easily clamped to a table, which makes this machine almost ready to go out of the box. But it's the accessories for this table that really make it shine.

An adjustable joint-making attachment (the Joint Maker, which comes standard) mounts to the rear of the table and supports the router horizontally (see the photo below). The Joint Maker is moved up and down with a crank. One full turn moves the bit vertically $1/16$ in.; it's easy to adjust to within $1/64$ in. or even $1/128$ in. This degree of precision makes it useful for cutting finger joints, dovetails, mortises and tenons, which is what it's designed for. The Joint Maker is also useful in a number of other routing situations. For example, with a safety cut, raised-panel bit, you can raise panels with the panels laying flat on the table instead of sticking up into the air, standing only on a thin edge.

The Porta-Nails Joint Maker accessory, which comes standard with the table, mounts the router horizontally. Used along with the miter gauge, it lets you rout tenons, mortises, finger joints and dovetails quickly and easily. It also makes other operations such as panel-raising safer and less prone to error.

Another standard feature is the safety starting pin. Porta-Nails router table is the only one besides the Porter-Cable Model 695 to offer this feature.

Options include a small miter gauge for $12.95 and a fence for $19.95. The miter gauge is necessary because the miter-gauge slot on this table is positioned too close to the bit for a standard miter gauge to pass by the bit. But the fence is just a simple piece of aluminum with a hole in it; for $19.95, you can do as well or better making your own fence of wood.

Fifty dollars will buy you the optional pin-router attachment, which bolts easily to the Porta-Nails Joint Maker. It's money well spent. Cranking the Joint Maker's lead screw raises and lowers the pin over the center of the router bit. The pin can also be adjusted in, out, left and right. Once you've aligned the pin and bit, all you do is attach a template to the blank you're routing and run the template against the pin. Template-routing makes duplication of irregularly shaped parts a straightforward affair, and the Porta-Nails pin router attachment makes template-routing affordable.

Conclusions

Any of these tables lets you cut tenons and half-laps quickly and precisely with a miter gauge, a significant advantage over conventional router tables. The Porter-Cable Model 695 and the Woodstock International W2000 Rebel are both priced to compete with the plywood and MDF tables. The Porta-Nails universal router system and the Porter-Cable (for most operations) require very little tweaking before use, whereas the NuCraft Tools Model 100 requires drilling and tapping for the plastic

SOURCES OF SUPPLY

NUCRAFT TOOLS, INC.

PO Box 87616, Canton, MI 48187-0616; (313) 981-4454

PORTA-NAILS, INC.

PO Box 1257, Wilmington, NC 28402; (800) 634-9281

PORTER-CABLE CORP.

4825 Highway 45 N., PO Box 2468, Jackson, TN 38302-2468; (901) 669-8600

WOODSTOCK INTERNATIONAL, INC.

PO Box 2027, Bellingham, WA 98227; (206) 734-3482

insert, and the Woodstock International plastic insert needs to have its corner holes and center hole drilled. The NuCraft Tools table, with or without legs, extension wing and shaper fence is without a doubt the toughest of the four, but it's also the heaviest and the most expensive ($685 with all the accessories listed here).

Each of the tables has its strengths and its weaknesses, but the table that really stands out is the Porta-Nails universal router system. At $458, including the miter gauge and the pin-router attachment, this table is a true bargain, letting you do far more than you can with a regular router table even with a miter gauge. We will most likely be seeing more systems like this one in the next couple of years as other manufacturers become aware of the router table's real potential.

STOW-AWAY ROUTER TABLE

by Jim Wright

Stow-away router table

*This router table clamps to your workbench in seconds.
Measurements are guidelines; size parts to fit your bench
for secure mounting.*

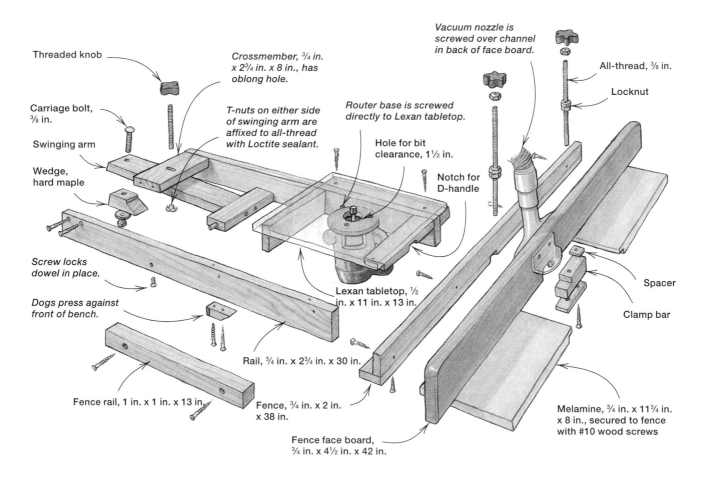

Threaded knob

Crossmember, ³⁄₄ in.
x 2³⁄₄ in. x 8 in., has
oblong hole.

Vacuum nozzle is
screwed over channel
in back of face board.

All-thread, ³⁄₈ in.

Locknut

Carriage bolt,
³⁄₈ in.

T-nuts on either side
of swinging arm are
affixed to all-thread
with Loctite sealant.

Router base is screwed
directly to Lexan tabletop.

Swinging arm

Hole for bit
clearance, 1¹⁄₂ in.

Wedge,
hard maple

Notch for
D-handle

Screw locks
dowel in place.

Spacer

Dogs press against
front of bench.

Clamp bar

Lexan tabletop, ¹⁄₂
in. x 11 in. x 13 in.

Rail, ³⁄₄ in. x 2³⁄₄ in. x 30 in.

Fence rail, 1 in. x 1 in. x 13 in.

Fence, ³⁄₄ in. x 2 in.
x 38 in.

Melamine, ³⁄₄ in. x 11³⁄₄ in.
x 8 in., secured to fence
with #10 wood screws

Fence face board,
³⁄₄ in. x 4¹⁄₂ in. x 42 in.

Stow-away router table details _____

BENCH-FRAME CONNECTION

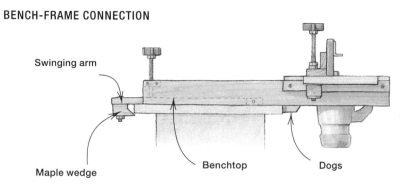

Swinging arm

Maple wedge

Benchtop

Dogs

ADJUSTABLE FENCE

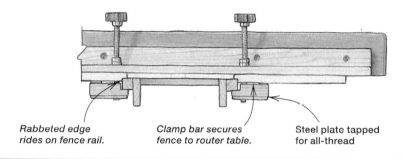

*Rabbeted edge
rides on fence rail.*

*Clamp bar secures
fence to router table.*

*Steel plate tapped
for all-thread*

I'm afraid my early attempts at making a router table were nothing to write home about. The first three designs—a table with legs, a cabinet base and a table attached to my tablesaw—all ended up on the scrap heap. They were just too bulky, a fatal flaw when it came to my small shop and lack of storage space.

It finally occurred to me all I really needed was a simple router table that could be clamped to my workbench. Here's how it works: I attach my router to a plastic insert and drop it into the frame of the table; the router hangs over the front edge of my bench. A sliding fence rides on top of the frame and adjusts easily. The mass of the bench kept the router table from vibrating, but best of all, the whole assembly is compact and easy to store (see the drawings on the facing page and above).

The key was in finding a way to clamp the assembly to my bench, so it wouldn't move. I did that by sizing the frame of the router table, so it spanned my bench exactly. Then I held the frame in place with a simple clamp made of a wooden wedge and a length of all-thread.

Building the frame for your bench and router

The first step is to decide how large a tabletop you need for your miter. Allow enough room for the knobs and handles on your router, and give yourself room to adjust the router when it's attached to the table. Just how much is enough depends on your router. I have a D-handle on my router, so I had to cut a relief in the framing to accommodate it. I made the frame for my router table of hard maple. The corners are fastened with #10 wood screws, so assembly is easy.

The router table stays in place because it grabs both the front and the back edge of my workbench. Attached to the rear of the router-table frame is a maple wedge that hooks over the back edge of my bench.

Big router-table per-formance in a benchtop package— This shop-built router table sets up in seconds and stores easily when not in use. Securely clamped to your benchtop, it can do most anything more conventional router tables can.

Screwed into the bottom of the frame near the front of the table are two dogs that press against the front edge of the bench. I added 80-grit sandpaper on the inside faces of these pieces to give them a better bite. When I tighten the knob at the back of the frame, the wedge pulls up against the bottom of the bench and locks the frame into place. Because the dogs on the front of the frame are tight against the front of the bench, my router table really can't go anywhere.

The assembly should hold securely even with the clamp knob a little loose—that's important. If the fit between the router table and your bench is sloppy and the clamping mechanism were to fail, the table would fall on the floor. Not a nice picture: router, work and fingers all mixed together and heading for the deck.

Making the tabletop
The thick table is a piece of $^{1}/_{2}$-in. Lexan from a dealer's scrap pile. The 11-in. by 13-in. piece cost me $10, but expect to pay more if you have a piece cut to size from stock. Lexan, a polycarbonate, cuts easily with a tablesaw, and trimming it to size is no problem. Other plastics may shatter or melt, but Lexan is lovely to work with. Phenolics also work well and are more rigid

than Lexan. (For more on plastics in the woodshop, see *FWW* #105 p. 58.)

The base of the router dictates the layout for the mounting holes. Use a drill press with a spade bit turning at low speed (clamp the work) and light pressure to cut a $1^{1}/_{2}$-in. hole in the center of the base (a piece of scrapwood beneath the work when drilling will keep the bit from jumping when it breaks through). The mounting holes are made with a twist bit, then countersunk. The Lexan is attached to the frame with four #10 wood screws.

Fabricating the fence
The fence is made to slide on the frame. Two clamps lock it in place. Attached to the right and the left sides of the fence are two pieces of melamine, which support the work as it's fed past the bit. The fence slides on or off in seconds. Once the router is mounted to the table, the table itself can be mounted or removed from the bench in 15 seconds— without changing the position of the fence.

The fence has a face board screwed to it with a channel in the back to create a duct for dust collection, as shown in the drawing. I added a plastic finger guard for safety. As a bonus, I find that the guard helps control the dust.

MORTISING WITH A ROUTER

by Gary Rogowski

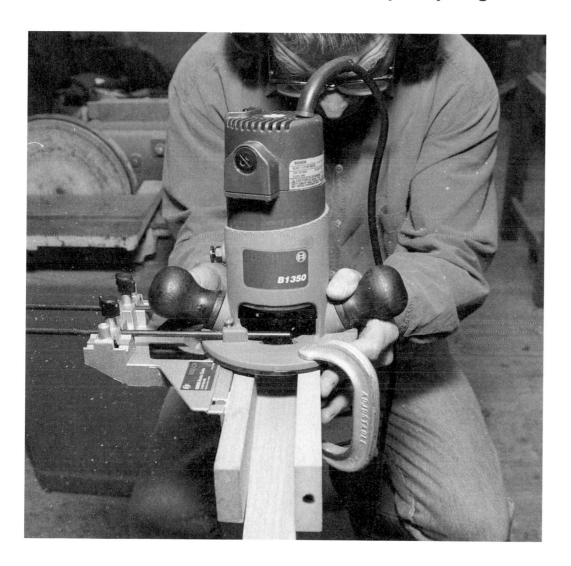

I cut my first set of mortises by hand. It was a fabulous learning experience. I found that chopping through red oak was like digging postholes in dry clay. I had to resharpen my chisel after each mortise, but I learned. I also bought a router.

A router is the quickest and most accurate tool for cutting mortises. Its versatility and speed is unmatched, and it can be used in a variety of setups, both upright and upside down in a router table. In minutes, a router cuts mortises that would take hours by hand. And you can reproduce your results with a minimum of hassle or setup time. When I have mortises to cut these days, the router is my first choice. Either a fixed-base or a plunge router can produce excellent results.

Choosing the right bit

There are a variety of bit sizes and types that can be used for mortising (for more on router bits, see *FWW* #116, pp. 44-48). Two shank sizes are commonly available: 1/4 in. and 1/2 in. Either will work, but bits with 1/2-in. shanks flex less under load, give a better cut and are less likely to break.

I don't bother with high-speed steel (HSS) bits because they need to be sharpened too often. Carbide-tipped bits cost two to three times more but they last much longer. Solid-carbide bits are great, too, but they're even more expensive.

Straight bits come in two flavors: single flute for quick removal of material and double flute for a smooth finish. Because you'll find double-fluted bits in most tool catalogs, you'll get more size options.

The flutes of a spiral bit twist around the shank. This gives a shearing cut that is even smoother than one from a double-fluted straight bit. Spiral bits are available both in solid carbide and carbide-tipped steel. They spiral up or down.

An up-cut spiral bit cuts quickly while pulling most of the chips out of the mortise. However, it also will tend to pull the work-piece up if it's not securely fastened. The up-cut spiral also can leave a slightly ragged edge at the top of the mortise where wood fibers are unsupported. Because the edges of a mortise are usually covered by the shoulders of a tenon, this kind of tearout generally isn't a problem.

A down-cut spiral bit pushes the work and the chips down. The result is a cleaner mortise but one that can become clogged with debris.

I have used mostly double-fluted straight bits and a carbide-tipped up-cut spiral bit. Recently, though, I bought a solid-carbide up-cut spiral, which cuts even better.

Using a fixed-base router

If the only router you have is a fixed-base router, you're not out of luck. It will just take a little more attention to detail and skill to get good mortises than it would with a plunge router.

A straight fence attached to the router is essential for accurately guiding the cut. Adding a long wooden auxiliary fence to your router's stock fence will give the router more stability. A second fence, clamped to the router base and on the other side of the workpiece, is a good idea, too (see the photo at left below). This fixes the position of the router laterally, so it can't accidentally slip to one side or the other during the cut. Combined with end stops, a double fence will virtually ensure accurately located mortises. The only thing left to set is the depth, and here you have a choice of methods.

Multiple depth settings

One way of mortising with a fixed-base router is to take just a little bite with each pass, gradually lowering the bit until you're at full depth. The biggest drawback with

Two fences keep a router in line. When routing to full depth with a fixed-base router, you want to make sure it doesn't veer out of the mortise.

Multiple depth settings create steps. To adjust the bit height on most fixed-base routers, you have to twist the motor in its base. This often results in stepped, sloppy sidewalls.

Mortising with a fixed-base router _____

You will get a cleaner mortise by setting the bit to full-depth right from the start. Cut the mortise in several passes with the router tipped at an angle.

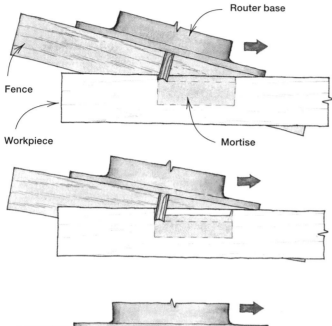

FIRST PASS
The first pass, with the router held at an angle, should remove about ⅛ in. to ¼ in. of material.

SECOND PASS
The angle of the router is lowered for the second pass, but bit depth remains the same. This and each successive pass removes from ⅛ in. to ¼ in.

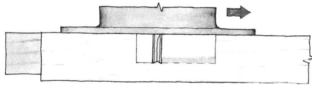

LAST PASS
The last pass is made with the router flat against the workpiece and the bit straight up and down.

Labels on figure: Router base, Fence, Workpiece, Mortise

---

this approach is that it's hard to get a smooth-walled mortise. The reason is that the motor and, consequently, the bit may not stay centered in the base as you adjust the depth of cut.

With most routers, adjusting the bit height requires that you turn the motor in the base housing. When you do, the bit moves in relation to the fence, only slightly, but enough to give the walls of the mortise a stepped, rough surface (see the photo at right on the facing page). Exceptions are DeWalt, Black & Decker and Elu routers, which employ a rack-and-pinion adjustment system that keeps the collet and bit centered at a fixed distance from the fence.

One depth setting

One way around this stepping problem is to set the bit at full depth right from the

start. To mortise, you just move the router at an angle to the workpiece so you introduce a little more of the bit to the wood with each pass (see the drawing above). The router is tilted, resting on one edge of its base, until the final pass is made. An extra-wide auxiliary fence is advisable, and a second fence clamped to the router base on the other side of the workpiece is essential.

Router-table mortises

Why would anyone want to cut mortises on a router table? Well, for narrower stock, a router table provides plenty of support. When routing narrow pieces from above, a hand-held router can become tippy and unstable. The edge of a door stile, for example, just doesn't offer very much support for a router base. With a router table, you have

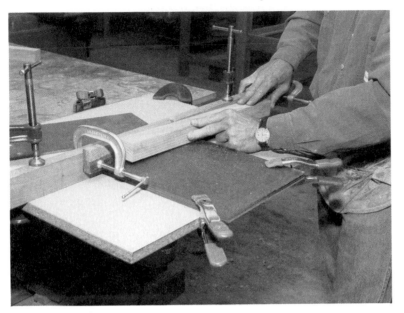

With hardboard shims, you set the bit just once. By removing one shim after each pass, you can take safe, manageable bites without having to change the router's depth setting. Increments of either ¼ in. or ⅛ in. are possible.

both the table and the fence against which to register the workpiece, and you only have the weight of the workpiece to control. For small table legs or cabinet doors, mortising on the router table is worth trying.

When mortising on a router table, use the fence to position the mortise from side to side and stops to establish the ends of the mortise. As you face the table, the work should move from right to left. This feed direction will help keep the work tight against the fence. Start with the workpiece against the right-hand stop, and lower the work into the bit. Because most bits don't cut in the center, it helps to lower and simultaneously move the work along just a little to avoid burning. Move the workpiece from right to left across the bit until it hits the other stop.

If you're using a plunge router, you can set the turret stops for incremental cuts, make three passes and finish up at full-depth. But if you're using a fixed-base router, you'll have a problem getting a smooth-walled mortise if you adjust the bit height between passes—just as you would when using the router upright.

My solution to this problem is to use shims made from ¼-in. hardboard, like Masonite, notched around the bit, to elevate the workpiece above the table (see the photo above). In this way, I can set the bit

at full height and just remove a shim after each cut, gradually working down until the workpiece is on the table and the last cut is made.

Mortising with a plunge router

The best tool for mortising is the plunge router used on top of the work. This is the job it was designed for. There are many different kinds of fixtures that can be used with the plunge router. Two that I use frequently, a U-shaped box and a template with a fence, are discussed below.

There are several schools of thought as to how to plunge the bit into the work. One method is to plunge a full-depth hole at each end of the mortise and then make a series of cleanup passes between those two holes. The drawback to this method is that you may get some burning as you plunge to full depth because most bits don't have center-cutting capability.

Alternately, you can make a series of successively deeper, full-length passes, always moving left to right with the bit lowered and locked in place each time. For me, making full passes without locking the plunge mechanism on each pass works best. I keep the router moving. Try each of these methods to see which one works best for you.

Using a stock router fence

The simplest method for mortising with a plunge router is to mark the mortise ends on the workpiece and to set the last turret stop for the full depth of the mortise.

To adjust the bit's position, place the router on a marked-out workpiece, and lower the bit so it's just touching the surface of the work. Rotate the bit so its cutting edges are in line with the width of the mortise. Adjust the fence so it's flush against the side of the workpiece and the edges of the bit are within the layout lines. Then clamp the workpiece firmly to the bench, and rout away. Keep in mind, though, that the router will be tippy on narrow stock.

You can try to bring the bit just up to the end marks of the mortise with each pass, but it can be difficult to see them with all those chips flying around. Another way to

Layout lines are much easier to see when they're not hidden. The pencil marks show where the router base should stop, and the tick marks indicate that you're getting close.

Stops are foolproof. Clamp or screw stops in place to limit the travel of the router, front and back. You won't have to worry about trying to see layout marks when the chips are flying.

accomplish this is to line up the edges of the bit at both ends of the mortise and make a pencil mark at the outside edge of the router base (see the left photo above). These marks are a lot easier to see than layout lines at the ends of the mortise.

If you're concerned about cutting beyond the layout lines, just clamp on stops to limit router travel. It only takes a second. Clamp the stops directly onto the workpiece once you've determined the length of the mortise (see the right photo above).

Mortising with the U-shaped box

One of the most versatile router-mortising fixtures that I've come across is a simple U-shaped box (see the drawing at right). I first saw one of these boxes in a magazine article by Tage Frid. Since then, I've made a number of them dedicated to particular pieces of furniture.

But having one fixture that handles a variety of different-sized parts is really useful, too. The one in the top photo on p. 62 is made of $3/4$-in.-thick medium-density fiberboard (MDF). Its sides are rabbeted for the bottom (this helps align it during assembly). I also made the bottom longer than the side walls so I could clamp it down easily to any work surface.

U-shaped mortising fixture

This router fixture is simple to make and incredibly versatile. It can be made to accommodate a wide range of work and only takes a few minutes to set up for a mortising operation.

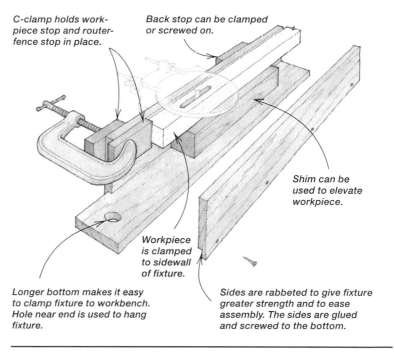

C-clamp holds workpiece stop and router-fence stop in place.

Back stop can be clamped or screwed on.

Shim can be used to elevate workpiece.

Workpiece is clamped to sidewall of fixture.

Longer bottom makes it easy to clamp fixture to workbench. Hole near end is used to hang fixture.

Sides are rabbeted to give fixture greater strength and to ease assembly. The sides are glued and screwed to the bottom.

One fixture cuts many mortises—This simple U-shaped box is one of the most versatile mortising fixtures you can build.

Measure once, and clamp a stop in place. The less measuring you have to do, the fewer errors you're likely to make. The stop on the inside of the fixture positions the workpiece. The one on the outside is a fence stop, which establishes one end of the mortise.

The best way to deal with multiple identical mortises is to clamp an end stop to the side wall of the fixture (see the photo at left). This way, each new piece will automatically be fixed in the right spot. Stops to index the mortise length also can be clamped onto the fixture. I prefer clamping these on rather than using an adjustable stop—I don't want to risk the stop being nudged out of place.

When placing the workpiece in the fixture, always make sure the piece is sitting flat on the fixture bottom and tight to the inside wall and end stop. Clamp the workpiece securely. Spacers can be used underneath pieces to bring them higher in the fixture or to push a piece away from the sidewall. Make sure the spacers are milled flat and support the workpiece well. Be sure that the clamps holding the work don't get in the way of the router.

To improve stability, attach a wooden auxiliary fence to the one that comes with the plunge router. Then position the bit in the right spot. Remember to hold the fence tightly to the wall, and be sure to move the router so the fence will be drawn up against the wall of the fixture by the rotation of the bit.

Dedicated mortising fixtures are extremely useful when you plan to reproduce a number of cuts on a regular basis. I made an angled fixture to cut the mortises for a stool I build at least once a year. The end stop locates each leg in the proper spot. A spacer block positioned against the stop locates the second set of mortises in the legs. Stops screwed to the outside of the fixture wall limit the length of travel of the fence and, therefore, produce mortises that are the correct length.

Templates and template guides

A template guide is a round metal plate with a thin-walled rub collar that extends out from its base (see the top photo at right). The guide is screwed to the router base, and a router bit fits through it without touching the inside wall of the collar. The outer wall of the rub collar is guided by a straight edge or template as the router cuts (see the bottom photo at right).

Templates that are made of hardboard, plywood or MDF include a slot to guide the rub collar as it makes the cut. The template is clamped to a workpiece with its slot centered over the mortise. I make up a template for a mortise when I'm doing a job I expect to repeat.

To make a template, nail a piece of $1/4$-in. hardboard about 5 in. wide and 10 in. long to a piece of wood approximately 2 in. sq. and a little longer than the hardboard (see the drawing on p. 64).

The wood block is the fence, and the hardboard gets a slot cut in it that is exactly the width of the rub collar. Cut the slot in the template on the router table. To be sure the slot is parallel with the fence—which ensures that the mortise is square to the stock you're routing—tack the hardboard back a little bit from the edge.

Set up the router table with a straight bit that matches the outside dimension of the

A template guide screws to the router base. The guide's rub collar follows a slot cut in the template.

Stops are built in. A template prevents side-to-side movement of the bit and automatically sets the length of the mortise.

rub collar. The template slot is pencil-marked on the hardboard. The diameter of the template guide is greater than that of the bit you'll use when mortising. So you'll need to add the distance from the outside of the rub collar to the edge of the router bit to each end of the slot in the template. Typically, this offset is between $1/16$ in. and $1/8$ in.

Before cutting the slot in the hardboard template, take a minute to determine the setback from the edge of the workpiece to the edge of the mortise. Then set the router-table fence accordingly. I like to double-check that the fence is in the right spot. So I make a nibble cut at the end of the template, and then measure the distance from that point to the fence. This method ensures that you get the correct distance. Once you have it, plunge the template down onto the

Using a template and guide to mortise _____

Routing mortises with a hardboard template and router-template guide is quick and virtually foolproof. The size of the slot in the hardboard determines the size of the mortise. The template is clamped to the workpiece, and the assembly is then clamped to the bench.

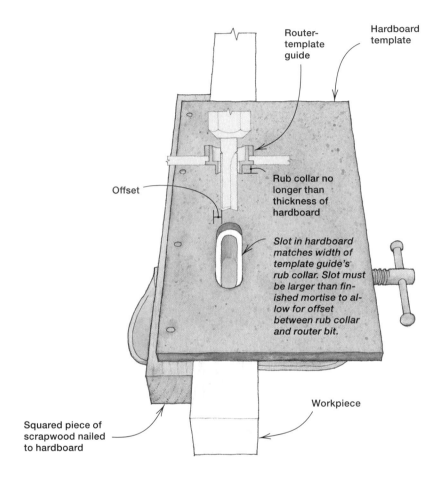

Router-template guide

Hardboard template

Offset

Rub collar no longer than thickness of hardboard

Slot in hardboard matches width of template guide's rub collar. Slot must be larger than finished mortise to allow for offset between rub collar and router bit.

Squared piece of scrapwood nailed to hardboard

Workpiece

---

bit as close to the center of the slot as possible, and then slide the template back and forth just up to the pencil marks at each end.

Templates like these are versatile. For example, a template made to cut a mortise $^3/_4$ in. from the edge of a table leg also could be used to cut the same sized mortise $^1/_2$ in. from the edge. How? Simply by inserting a $^1/_4$-in. shim between the template fence and the workpiece.

Once you have made the template and clamped it to the workpiece, position the plunge router with the template guide on the work. Set the bit depth, taking into account the thickness of the template. An up-cut spiral bit will pull most of the debris out of the mortise as the cut is made. Compressed air can help clear a mortise that's really packed with chips.

SIMPLE FIXTURE FOR ROUTING TENONS AND MORE

by Patrick Warner

Routing or shaping the end of a board can be a tricky proposition. Even on a router table fitted with a fence, the small amount of surface area on the end of a board doesn't provide much stability when you try to run the piece vertically past the bit. And if the stock is very long, the task is simply impossible because of the difficulty of handling a long piece on end, even if your shop's ceiling is high enough to allow it. My router end-work fixture provides a safe and simple solution for routing tenons as well as other joints or shapes on the end of a board.

How the fixture works

Basically, this is the way the fixture works: A frame member or other workpiece is clamped to the fixture, which references it for the desired cut. The fixture's large platform top provides a stable support for the hand-held router, and a window cutout in the platform allows access for the bit to shape the narrow end of the workpiece, as shown in the photo above. The fixture features an indexing fence that's adjustable to facilitate angled tenons, such as those used to join seat rails to the rear leg of a chair. The method of guiding the bit depends on the job. Some joints, such as stub tenons, can be done with a piloted rabbeting bit that rides the faces of the stock. An auxiliary router fence can be used to create more complicated tenons, sliding dovetails or other shapes on the end of stock, including roundovers or chamfers. The stock can be

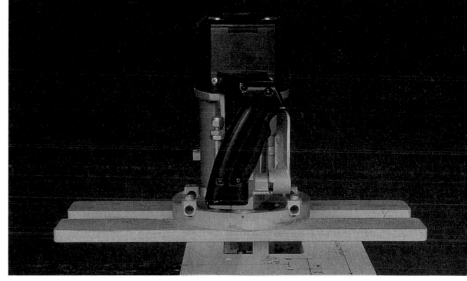

An end-work platform holds workpieces vertically for cutting tenons or rounding the ends of frame members with a router. The slats attached to the bottom of the router keep it from tipping as the router passes over the window in the top of the fixture.

any shape—square, rectangular or even round, as shown in the photo on page 66. Practically any bit normally used on the edge or face of a board can be used with this fixture. Because the router bit slices the wood fibers parallel to the grain when shaping the end of a board, the fibers are effortlessly peeled away rather than sheared, as is the case with cross-grain router cuts.

Building the fixture

The parts for the fixture can be made from any hardwood; beech, maple and birch are good choices (I built mine from birch), or you can use a good-grade of $3/4$-in. or 1-in. medium-density fiberboard (MDF). The fixture consists of a router platform with a rectangular window cutout for the router bit; a workpiece clamping board, joined at 90° to the platform with a tongue and groove and reinforced by two corner braces; and an adjustable indexing fence (see the drawing below). After cutting the platform to size, I rout a $^3/_8$-in.-wide by $^3/_{16}$-in.-deep groove the length of the bottom surface to receive the mating tongue on the clamping board. I rout the groove by running the router's accessory fence along the edge of the platform to ensure that the face of the clamping board will be parallel to the edge. Next, I cut out the window slightly undersized with a sabersaw. Then I trim it to final size with a router and a flush trimming bit following a template. A $3^3/_4$-in. by 6-in.

window allows routing on stock up to about 2 in. by 4 in. with bits up to $1^1/_2$ in. dia. If larger stock or bigger cutters are used, make the window and/or platform larger.

After cutting the clamping board to size and rabbeting its top edges to form the tongue, I cut out a portion of the top edge for router bit clearance when the fixture is put to work. I rough out the cut with a sabersaw and trim it using one side of the same template I used for the platform window. The $1^3/_{16}$-in. by $6^1/_4$-in. cutout in the drawing allows for routing workpieces to a depth of about $1^7/_8$ in. (if you take deeper cuts, make the cutout deeper, too). I bandsaw the corner braces from $^{15}/_{16}$-in.-thick pieces about $4^1/_4$ in. sq.

I use the tongue-and-groove joint to accurately register the clamping board to the platform, but I screw all the parts together instead of gluing them, so it's easier to disassemble and realign them later if necessary. To ensure that the screw holes in the platform align perfectly with those in the clamping board, I first drill four $^{13}/_{64}$-in.-dia. holes (for #10 flat-head screws) in the platform—two on either side of the window and centered on the groove. Then I use a $^{13}/_{64}$-in.-dia. transfer punch to mark the pattern for the pilot holes from the platform to the clamping board (see the sidebar on the facing page). I also use the same punch along with a countersink to perfectly prepare the holes for the heads of #10 flat-head screws (described in the sidebar). I use the same transfer and drilling process for drilling pilot holes in the corner braces.

Router end-work fixture

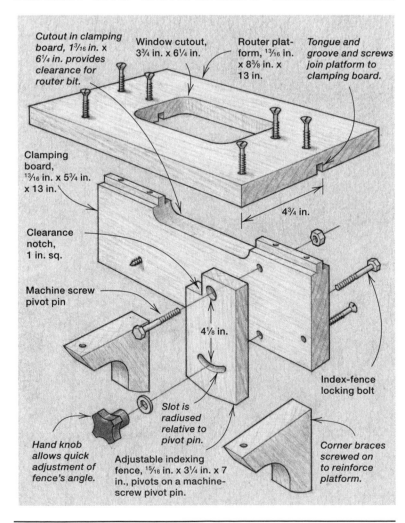

Cutout in clamping board, $1^3/_{16}$ in. x $6^1/_4$ in. provides clearance for router bit.

Window cutout, $3^3/_4$ in. x $6^1/_4$ in.

Router platform, $1^3/_{16}$ in. x $8^3/_8$ in. x 13 in.

Tongue and groove and screws join platform to clamping board.

Clamping board, $1^3/_{16}$ in. x $5^3/_4$ in. x 13 in.

4$^3/_4$ in.

Clearance notch, 1 in. sq.

Machine screw pivot pin

4$^1/_8$ in.

Index-fence locking bolt

Hand knob allows quick adjustment of fence's angle.

Slot is radiused relative to pivot pin.

Adjustable indexing fence, $^{15}/_{16}$ in. x $3^1/_4$ in. x 7 in., pivots on a machine-screw pivot pin.

Corner braces screwed on to reinforce platform.

A variety of joints can be routed with the end-work platform, including all kinds of square or angled tenons and sliding dovetails. Tenons can even be routed on the ends of round stock.

Machinist's transfer punches find a niche in the woodshop

Transfer punches are steel rods used to accurately mark the location of holes from an already drilled part to one that will be drilled to match. While they are tools from the machinist's chest, woodworkers can make good use of them as well. Typical jobs where transfer punches come in handy include drilling holes in a new router subbase using the old subbase as a pattern; locating and screwing a plinth or cornice to a carcase; and drilling pilot holes in jig parts that must fit accurately together (see the main article). While any of these jobs can be accomplished with a scratch awl, using a transfer punch is much more accurate.

Sets of transfer punches are sold in either standard fractional or metric sizes as well as special drill letter and number sizes. I purchased my set, as shown in the photo above, for

A set of machinist's transfer punches is a worthwhile investment for any woodshop. Not only do the precisely dimensioned punches provide a great way to transfer hole positions between parts, they can also be used for other drill-press jobs, such as centering previously drilled holes.

about $15 from Enco Manufacturing Co. (5000 W. Bloomingdale, Chicago, Ill. 60639), but they are also available at any good machinist supply house. Each punch is a few thousandths smaller in diameter than its corresponding drill size, so it's easy to slide in and out of an already drilled hole. The end of each rod is turned to a point so that it will put a dimple exactly in the center. To use a punch, first clamp the already drilled part in position over the part to be marked, insert the punch into the hole and lightly tap with a hammer. The slightly indented punch mark creates a starting dimple for the drill. Drill the new holes with a brad point bit, and you'll be amazed at the accuracy.

Other uses

While punches excel at transferring hole positions, there are other uses for them in the

woodshop. When drilling for flat-head wood screws, I often use an 82° countersink designed to be locked onto a drill bit with setscrews (available from W.L. Fuller, Inc., PO Box 8767, Warwick, R.I. 02888; 401-467-2900). I've found that these countersinks work better when mounted on a transfer punch, as shown in the photo below. Using a transfer punch instead of a drill as a pilot has two advantages: The unfluted punch doesn't tear up the hole, and the countersink stays cooler because the smooth punch doesn't trap the chips produced by the countersink's cutters (as a drill bit does).

A transfer punch also can be used to center a previously drilled hole on the drill press either to counterbore it or to increase its depth or diameter. First tighten the appropriate-sized punch in the chuck, then lower it into the hole and lock the drill-press quill.

Now you can clamp the part to the drill-press table, unlock the quill, insert the new bit and rebore as desired. A punch chucked in the drill press can also be used with a machinist's square to check the drill-press table for squareness to the bit. This same method also works to check square between a router's collet and a baseplate. A punch can be inserted into any hole, and its angle to the work surface can be checked with a machinist's square.

Finally, the rods can be used as form-sanding cauls for small coves. Because the punches come in almost any small diameter (less than $9/16$ in.), the thickness of the abrasive can be compensated for, and a perfect fit obtained, by using a smaller punch than the desired cove. Also, each set of transfer punches comes in a holder, and the holes in these holders can be used as a drill gauge for those drill bits that have lost their identity. —P.W.

A transfer punch can be used with a countersink mounted on it to cleanly and accurately prepare previously drilled holes for flat-head screws.

Adustable fence

Now I saw out the adjustable fence and cut out a 1-in.-sq. clearance notch from one corner, as shown in the drawing. I drill a $^1/_4$ in.-dia. hole, centered $^3/_4$ in. from the top end of the fence, for a pivot pin. Then I fit my plunge router with a $^1/_4$-in.-dia. straight bit and a circle cutting jig for routing a curved slot in the fence. This curved slot, which is centered on the fence about $4^1/_8$ in. from the pivot pin, allows the fence to be pivoted side to side and set either square to the platform or askew for angled tenons. Next, I clamp the fence to the clamping board so that the fence's corner notch is aligned with the clamping board's clearance cutout, and its top edge is about $^1/_8$ in. below the platform. Then I use a $^1/_4$-in.-dia. transfer punch to accurately mark the fence's pivot hole and the center of the curved slot on the clamping board. These hole centers have to be precise, or the fence won't adjust easily. After drilling both holes with a $^7/_{32}$-in. drill bit, I thread the holes in the hardwood with a $^1/_4$-20 tap and install a $1^3/_4$-in.-long, $^1/_4$-20 machine screw for the pivot pin and a $1^1/_2$-in.-long $^1/_4$-20 flat-head machine screw for the fence locking bolt. A threaded hand knob on the locking bolt makes fast fence adjustments without a wrench.

End routing stock

To use the fixture for routing basic tenons, first set the indexing fence precisely 90° to the router platform. Now, set the fixture upside down on the bench, position the stock to be tenoned against the indexing fence with the stock's end flat on the bench, and secure it to the clamping board with a couple of C-clamps (see the photo at left below). This indexes the workpiece square to, and flush with, the top surface of the platform. Flip the entire assembly over and clamp the workpiece in the bench vise so that the router platform is at a comfortable working height.

To eliminate any chance that the router will tip as it passes over the window in the fixture's platform, I screw a couple of $^1/_2$-in.-thick strips of wood to the router base. You also could cut out and screw on an oversized subbase, made from Masonite or Plexiglas. If the desired cut can be made at a single pass, such as for a stub tenon, any standard router will do. Simply chuck up a piloted bit, set the cutting depth (which determines the tenon's length) and guide the bit around the stock (see the photo at right below). Rabbet bits and pilot bearings of various diameters can be mixed or matched to produce tenons with shoulders from $^1/_{16}$ in. wide to $^9/_{16}$ in. wide. Fit the router with an auxiliary guide that runs along the platform's edge when unpiloted cutters are used.

For deep cuts, like tenons that are longer than the cutting depth of the bit, a plunge router is my tool of choice. I set my plunge router's rotary depth stop to three different cutting heights and then shape each tenon in three passes, resetting the stop to take a deeper cut each time. By changing bits and cutting heights, tenon shoulders can be cut at different heights, centered or offset.

After setting the adjustable indexing fence for shaping either a square or angled tenon, the fixture is held upside down on a flat surface, and the workpiece is clamped in place with its end flush with the top of the router platform. The fixture is then flipped over and clamped in a bench vise for routing.

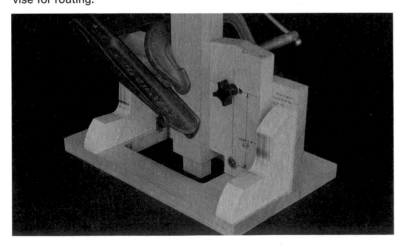

Shaping tenons with a piloted rabbet bit is simple: The pilot bearing rides on the face of the workpiece as the short tenon is cut.

ROUTER FIXTURE TAKES ON ANGLED TENONS

by Edward Koizumi

We live in a turn-of-the-century Arts-and-Crafts house, so it seemed quite natural to furnish it with pieces from that era. My wife bought a pair of Mission armchairs a couple of years ago to go with a 9-ft.-long cherry table I'd built for our dining room. Six months later, she bought two side chairs. It would be a while before we could afford a full set. Within earshot of my wife, I heard myself say, "How hard could it be to make these?"

"Oh, could you?" she asked.

"Sure," I said. The chairs looked straightforward enough, just a cube with a back. Upon closer examination, I realized that the seat was slightly higher and wider in the front than in the back. For the first time, I was faced with compound-angled joinery. I thought about dowels, biscuits and loose tenons, so I could keep the joinery simple, but I wasn't confident in the strength or longevity of these methods.

I wanted good, old-fashioned, dependable mortise-and-tenon joints. After some thought, I decided an adjustable router fixture would be the simplest solution that would let me make tenons of widely varying sizes and angles (see the photos at right).

The fixture I came up with is as easy to set up as a tablesaw. In fact, there are some similarities (see the drawing on p. 70). The workpiece is held below a tabletop in a trunnion-type assembly that adjusts the tilt angle (see the bottom photo at right). For compound angles, a miter bar rotates the workpiece in the other plane. The fixture can handle stock up to 2 in. thick and 5 in. wide (at 0°-0°) and angles up to 25° in one

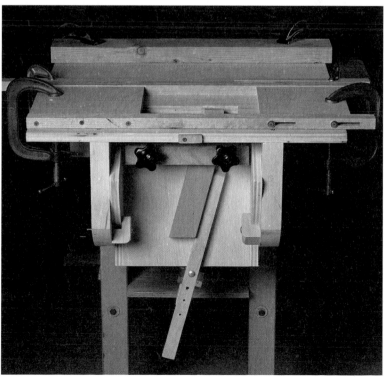

Tenon-routing fixture for compound angles

This fixture, adjustable in two planes, is designed to let you rout compound-angled tenons consistently and accurately. The tenons can be either squared or rounded, depending on which guide frame you use.

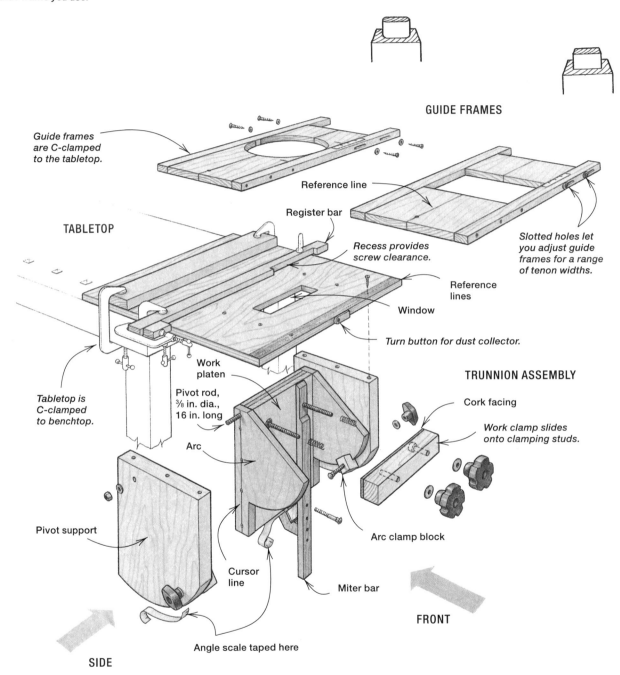

GUIDE FRAMES

Guide frames are C-clamped to the tabletop.

Reference line

Register bar

TABLETOP

Recess provides screw clearance.

Reference lines

Window

Slotted holes let you adjust guide frames for a range of tenon widths.

Turn button for dust collector.

Tabletop is C-clamped to benchtop.

Work platen

Pivot rod, ⅜ in. dia., 16 in. long

Arc

TRUNNION ASSEMBLY

Cork facing

Work clamp slides onto clamping studs.

Pivot support

Cursor line

Arc clamp block

Miter bar

FRONT

Angle scale taped here

SIDE

plane and 20° in the other. This is sufficient for chairs, which seldom have angles more than 5°.

To guide the router during the cut, I clamp a guide frame to the fixture over the window in the tabletop (more on positioning it later). And I plunge rout around the tenon on the end of the workpiece. The guide frame determines the tenon's width and length, as well as whether the ends will be square or round (see the top photo on p. 69). I made two frames, both adjustable, one for round-cornered tenons, the other for square tenons.

The fixture and guide frames took me just over a day to make, once I'd figured out the design. Then I spent about an hour aligning

Guide frame determines thickness and width of tenons. The author keeps the router's base against the inner edges of the guide frame and routs clockwise to prevent tearout. Guide frames can produce round-cornered or square-cornered tenons.

Set correctly, the fixture will yield tight joints, whether the tenons are straight, angled or compound-angled. Here, the author tests the fit of a seat-rail tenon into a leg mortise.

the fixture and making test tenons in preparation for routing the tenons on the chair parts. The fixture worked just as planned and allowed this relatively inexperienced woodworker to produce eight chairs that match the originals perfectly.

Making the fixture and guide frames
The fixture is simple to build. It consists of only two main parts, the trunnion assembly and the tabletop. The trunnion assembly (see the drawing on p. 70) is essentially a pair of arcs nestled between two pivot supports. Between the two arcs is a work platen, or surface, against which I clamp the component to be tenoned. There are other parts, but basically, the fixture is just a table to slide the router on and a movable platen to mount the workpiece on.

I built the fixture from the inside out, beginning with the work platen (see the drawing on p. 70). Because I didn't have any means of boring a 10-in.-long hole for the threaded rod on which the arcs pivot, I dadoed a slot in the platen and then glued

in a filler strip. Next I located, center punched and drilled the holes for the T-nuts and retaining nuts that hold the clamping studs in place. Center punching ensures that the holes are exactly where they're supposed to be, which is important for a fixture that's going to be used over and over again. I center-punched the location for every hole in this fixture before drilling.

Before attaching the clamping studs to the work platen, I made the arcs, which go on the sides of the work platen. I laid out the arcs (and the pivot supports) with a compass, bandsawed and sanded the arcs, and drilled a hole for the pivot rod through the pair. I glued and screwed the arcs to the platen. After giving the glue an hour or so to set, I tapped the T-nuts into the back of the work platen, screwed in the clamping studs and twisted on retaining nuts, which I tightened with a socket and a pair of pliers.

I made the pivot supports next. Then I cut a piece of threaded rod 16 in. long and deburred its ends with a mill file. I slipped the threaded rod through the pivot supports,

Angles in one plane (side view) ____

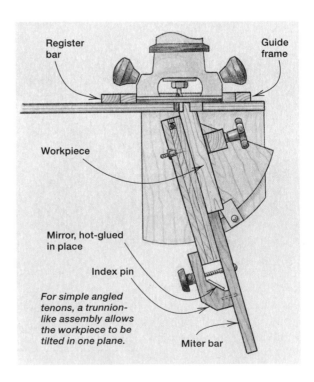

Register bar

Guide frame

Workpiece

Mirror, hot-glued in place

Index pin

For simple angled tenons, a trunnion-like assembly allows the workpiece to be tilted in one plane.

Miter bar

arcs and work platen, capped it at both ends with a nut and washer, and made and attached the arc clamps (see the drawing on the facing page).

Then came the tabletop. I cut it to size, cut a window in it and marked reference lines every 1/8 in. along the front edge for the first 2 in. With the tabletop upside down on a pair of sawhorses, I put the trunnion assembly upside down on the underside of the tabletop. Then I positioned the front of the pivot supports against the front edge of the tabletop and made sure the work platen was precisely parallel to the front edge and centered left to right. That done, I drilled and countersunk holes for connecting screws through the tabletop into the pivot supports. I glued and screwed the pivot supports to the tabletop.

Then it was time to make the miter bar, miter-bar clamp and the work clamp (see the drawings on the facing page and below). The mirror on the miter-bar clamp makes it easy to read the angle scale from above. I faced the work clamp with cork to prevent marring workpieces and counterbored it to take up the release springs. The release springs are a nice touch. They exert a slight outward pressure on the work clamp, causing it to move away from the platen when loosening the knobs to remove a workpiece.

The guide frames

Now for the guide frames, which clamp to the tabletop and limit the travel of the router. I made the frames adjustable lengthwise to handle a variety of tenoning situations. But their width is fixed. To determine the width of the frames, I added together the desired tenon width, the diameter of the bit I was using and the diameter of the router base. If your plunge router doesn't have a round base, you should either make one from acrylic or polycarbonate (you can cut it with a circle-cutting jig on a bandsaw), or buy an aftermarket version. I screwed the frame together in case I need to alter the opening later (for a new router bit, for example). I marked a centerline along the length of the frame on both ends.

Compound angles (front view) _____

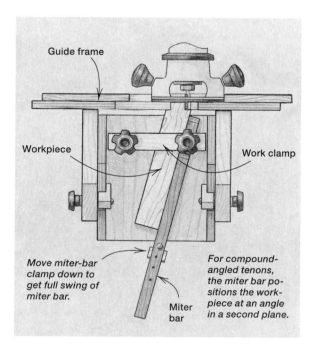

Guide frame

Workpiece

Work clamp

Move miter-bar clamp down to get full swing of miter bar.

Miter bar

For compound-angled tenons, the miter bar positions the workpiece at an angle in a second plane.

Initial alignment

Before I could use the fixture, I had to get everything in proper alignment and put some angle scales on it. I printed out some angle scales from my personal computer and taped them to my fixture with double-faced tape. But a protractor and bevel gauge also will work just fine to create angle scales for both the tilt angle and the miter angle.

To align the parts of the fixture, I flipped it upside down on the end of my bench and clamped it there. I used a framing square to set both the work platen and the miter bar at 90°, sticking the blade of the square up through the window of the tabletop and resting the tongue of the square flush against the inverted face of the tabletop. Then I stuck the angle scales on the two pivot supports and on the bottom of the work platen.

Routing test tenons

Next I routed test tenons with the fixture set at 0°-0°. I positioned the guide frame parallel to the front edge and centered on the window in the tabletop and clamped it to the fixture. I clamped a test piece the same thickness and width as the actual component in the fixture, with one end flush with the top surface of the tabletop. To do this,

I brought the test piece up so that it just touched a flat bar lying across the window. I set my plunge router for the correct depth and routed the tenon clockwise to prevent tearout.

I made a test mortise using the same bit I planned to use for the mortises in the chair. The fit wasn't quite right. So I adjusted and shimmed the frame until the tenon fit perfectly. If you rout away too much material and end up with a sloppy tenon on your test piece, you can just lop off the end and start over.

Once I had a tenon that was dead-on, I made an acetate pattern that allowed me to position the guide frame accurately for all tenons of the same size, regardless of the angle. I cut a heavy sheet of acetate (available at most art-supply stores) so that it would just fit into the guide-frame opening. I marked a centerline along the length of the acetate that lines up with the centerline down both ends of the guide frames. I also indicated which end was up and where the acetate registered against the guide frame. Then I put the test piece with the perfectly fitted tenon back into the fixture, laid the acetate into the opening in the guide frame and traced around the perimeter of the tenon end using a fine-tip permanent marker.

Setting up for angled tenons—Mark out the tenon on a test piece. The test piece should be the same thickness and width as the actual components, but length isn't important.

Make the workpiece flush with the tabletop. The author uses a piece of milled steel, but the edge of a 6-in. ruler would work as well.

Routing angled tenons

With the pattern, routing angled tenons is pretty straightforward. I crosscut the ends of all the pieces I was tenoning at the appropriate angles and marked out the first tenon of each type on two adjacent sides, taking the angles off a set of full-scale plans. Then I extended the lines up and across the end of the workpiece (see the photo at left on the facing page).

Having set the fixture to the correct angles, I brought the workpiece flush with the tabletop using a flat piece of steel as a reference (see the photo at right on the facing page). Then I clamped the workpiece in place. Finally, I set the acetate pattern in the guide-frame opening and positioned the guide frame so that the pattern and the marked tenon were perfectly aligned (see the photo below). With the guide frame clamped in place, I removed the acetate and routed that tenon. All other identical tenons needed only to be flushed up and routed. After the first, it was quick work.

There are pitfalls though. I found it important to chalk orientation marks on each workpiece. It can get confusing with two angles, each with two possible directions. And I had to be especially careful when routing the second end of a compo-nent. Make sure it's oriented correctly relative to the first. I messed up a couple of times and have learned to plan for mistakes by milling extra parts and test pieces. You might even end up with an extra chair.

To get flat surfaces on curved parts so I could clamp them in the fixture, I saved the complementary offcuts and taped them to the piece I was tenoning. Or I could have tenoned first and bandsawed the curves later.

For pieces with shoulders wider than the bit I'm using to remove waste, I clamp a straight piece of wood—a register bar—against the guide frame (a small pocket for screw clearance may need to be made), as shown in the drawing on p. 70. That way I can rout most of the tenon, unclamp the guide frame, slide it forward (using the reference lines at the forward end of the table-top to keep it parallel), clamp it down and then rout the remainder. I start the next piece in the same place and return the guide frame to the original position to finish the tenon.

Make a pattern. An outline of the tenon traced on acetate helps align the guide frame for cutting any tenons of the same size.

MAKE YOUR OWN DOVETAIL JIG

by William H. Page

The blanket chest I wanted to make for a gift was basically a large box joined with dovetails at the corners. I didn't have enough time to hand-cut the joints and didn't want to pay $300 for a commercial jig to do the job, so I set to work developing my own jig.

Shop-built from scraps, these unusual jigs, one for the tails and one for the pins, cut tight-fitting through-dovetails (see the inset photo below), a task that even many commercial jigs can't handle. Designed for routing dovetails for large carcase construction, the jigs can be built in less than two hours for just pennies.

Layout is quite simple and can be done as the tail jig is being assembled. Fingers screwed to the tail jig guide the router bits; the key is ball bearings. The bits used to cut the joint are guided by bearings the same diameter as the cutter. Pin and tail size and spacing are variable, and jigs can be built to handle any width board.

Basics of jig construction

Before making any of the jigs, the project stock must be jointed, planed and cut to final dimensions. The stock should be flat and square, and be sure to include a couple

Precise through-dovetail joints (see inset photo) are easy to rout with the aid of a couple of shop-built jigs. Here, the author completes the second part of the joint by routing the pins with a bearing-guided straight bit. The bearing rides against pin templates that have been positioned accurately using a guide board routed with the tail jig, which is the first jig to be built.

Making dovetail jigs

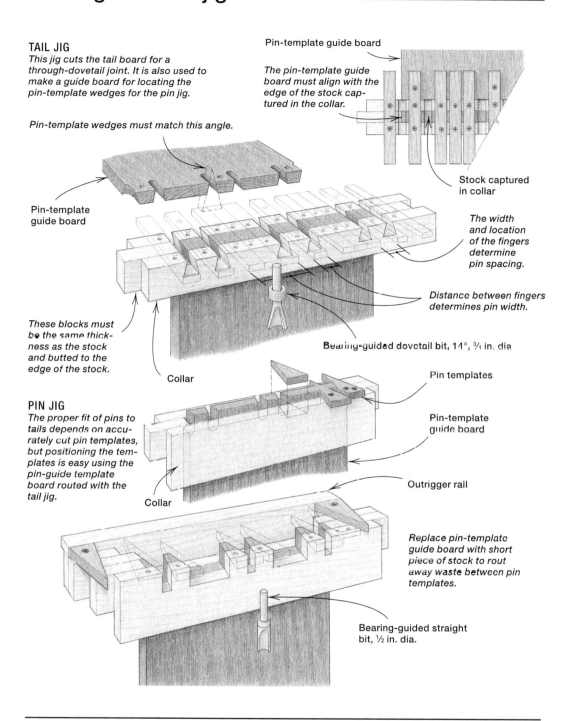

TAIL JIG

This jig cuts the tail board for a through-dovetail joint. It is also used to make a guide board for locating the pin-template wedges for the pin jig.

Pin-template wedges must match this angle.

Pin-template guide board

These blocks must be the same thickness as the stock and butted to the edge of the stock.

Collar

Pin-template guide board

The pin-template guide board must align with the edge of the stock captured in the collar.

Stock captured in collar

The width and location of the fingers determine pin spacing.

Distance between fingers determines pin width.

Bearing-guided dovetail bit, 14°, ¾ in. dia

PIN JIG

The proper fit of pins to tails depends on accurately cut pin templates, but positioning the templates is easy using the pin-guide template board routed with the tail jig.

Collar

Pin templates

Pin-template guide board

Outrigger rail

Replace pin-template guide board with short piece of stock to rout away waste between pin templates.

Bearing-guided straight bit, ½ in. dia.

extra feet of stock for making and testing the jigs. The jigs are assembled around some scraps cut from the actual stock. This way, the jigs precisely fit the stock and eliminate the need to fiddle with adjustments or set-up routines, ensuring perfect-fitting dovetails.

I start with the tail jig, and in the process of making this jig, I also cut a guide board that precisely locates the pin templates for assembling the pin jig. Using the tail jig to rout the pin-template guide board ensures a perfect match of pins to tails.

The tail jig consists of a collar that surrounds the stock to be joined and a series of fingers screwed to the top of this collar, as shown in the drawing on p. 77. The fingers serve as a stop when inserting stock into the collar and as a guide for the bearing on the bit. The location of these fingers across the top of the collar determines the spacing of the pins.

With the fingers in place, I run the dovetail bit through the collar of the jig and a scrap piece of stock clamped in the jig. These cuts create the tail piece, or the openings that the pins will fit into, and prepare the jig for use. The collar must be clamped to the stock to avoid any movement that could affect the accuracy of the cut.

The pin jig consists of a collar built around the pin-template guide board, but instead of straight fingers, the pin templates for this jig are wedges with an included angle to match the cut of the dovetail bit. An outrigger attached to the pin collar provides full support for the router when routing the pins.

With both jigs assembled, I rout a joint in a couple of pieces of scrap clamped firmly in the collars to test the fit and to be sure I like the pattern before proceeding with my good stock.

Making the tail jig

To make the tail jig, I clamp a short piece of the prepared stock in my bench vise. I begin by building the collar around this piece of stock, using 2-in.- to 3-in.-wide scraps that are about 4 in. longer than the width of the stock. The collar pieces are clamped flush to the end of the stock so that they overhang equally on both sides of the stock. The end collar blocks should be butted tightly to the side of the stock.

The guide fingers that are glued and screwed to the top of the collar are simply strips of hardwood or plywood about ⅝ in. thick and about 8 in. long. Position the strips for any pin pattern that you want, but keep in mind that the pins must be at least ¾ in. wide, the diameter of the bearing that will ride against the fingers. Also, the distance between the pins must be at least equal to the diameter of the straight bit used to cut the pins. The fingers also must be square to the collar. To avoid pin cutouts where I don't want them, I fill gaps between the fingers.

To make a tail jig, build a holding fixture that forms a collar around the stock, screw guide fingers to the top edge and then rout between the fingers to create the sockets.

To rout the tail jig, I chuck my bearing-equipped dovetail bit in the router and set the depth of cut about $1/32$ in. deeper than the thickness of my prepared stock. (Bearing guided bits can be made by gluing a bearing the same size as the bit to the bit's shaft, or they can be ordered from Freud, 218 Feld Ave., High Point, N.C. 27264.) I then rout the tails by running the bearing between the fingers. I make two passes in each slot to be sure the bearing rides firmly against both fingers for each pin cutout or else the pins won't align properly with the tails. This completes the tail jig, and in the process, I've made a scrap tail piece to test the fit of the joint.

Making the pin jig

The first step in making the pin jig is to use the tail jig for cutting the guide that locates the pin templates, so the pins and tails line up. I do this by butting a piece of prepared stock against the back side of the tail collar and screwing down through the fingers and into the jig stock. This piece of stock must be the same width as the workpieces to be joined, and its edges must align with the edges of the jig. To create the pin guide, I run my router between the fingers of the tail jig as before, cutting approximately an inch into the stock, as shown in the photo on the facing page. After routing, I unscrew the pin guide and then clamp it in my bench vise with the routed end up.

As with the tail jig, I build a four-piece collar around the pin guide clamped in the vise. I let the pin guide extend about $1/2$ in. above the collar, so the routed slots can be used to position the pin templates on the collar.

The pin templates are $5/8$-in.-thick wedges that I cut on the tablesaw from a long strip about 3 in. wide. I set my miter gauge to 14° (because I used a 14° dovetail bit), cut one edge of the wedge, flip the strip over and then cut the other edge of the wedge. I test-fit the wedge into the pin guide, make

minor adjustments to the miter gauge as necessary and then cut a new wedge. I continue this process until I get a wedge that fits snugly into the pin guide with no gaps on either edge. Then I cut a wedge, or pin template, for each slot in the pin guide.

I then push the pin templates firmly into place on the pin guide and glue and screw the templates to the collar. To fully support the router, I needed to attach an outrigger rail to the collar in front of the pointed end of the templates.

To rout the pins, I set up a second router with a bearing-guided, $1/2$-in.-dia. straight bit. Again, the depth of cut is just a hair deeper than the thickness of the stock. Before routing away the waste between the pins, I removed the pin-template guide board from the jig and replaced it with a short piece of the prepared stock. Then, with the router sitting on the pin templates and the outrigger rail, I routed away all the material on the collars and the scrap stock that is not covered by the pin templates. I used firm pressure to be sure the bearing rode tightly against the templates for an accurate cut. Routing the waste completes the jig and cuts a pin test piece.

I remove the test piece and try it in the tails previously cut. If the joint is too loose or too tight, it's usually a result of not keeping the guide bearing firmly against the sides of the fingers or pin templates. You might want to try running the router through the jigs again with new test pieces in place. Minor misfits can be adjusted by shaving the edge of the pin templates or adding masking-tape shims. If you're satisfied with the fit and spacing, slide the appropriate jig over the end of your stock and start cutting. The actual routing of joints takes about five minutes each.

TURN A ROUTER INTO A JOINT-MAKING MACHINE

by Guy Perez

I often turn to my router for joinery tasks. With a fence and straight bit, the router makes quick work of mortise-and-tenon joints. And for most projects, the consequent problem of either rounding the tenon or squaring the mortise is relatively minor. However, I recently made a crib with

44 slats that required 88 mortise-and-tenon joints. This daunting task prompted me to build my own version of the fancy joint-making machines I had often admired but had never been able to afford. I based the jig around my joinery needs and the funds I had to work with. And I confess that I

Guy Perez routs a tenon using a shop-built jig. Unlike jigs that require an operator to move control levers and machine, Perez's jig fixes the work, which lets him rout with two hands. The back of the jig (inset) reveals a carriage that guides both vertical travel (rods through bushings) and horizontal motion (rails captured by bearings). The router's plunge mechanism sets depth of cut.

arrived at much of the design during some less-than-inspirational philosophy seminars.

The jig I built operates somewhat like the commercially available machines, such as the Matchmaker or the Multi-Router, which cost from around $600 to over $1,500. With my jig, the work stays fixed, which means I can move the router with two hands (see the photo at left on the facing page) instead of having to manipulate the workpiece, router and control levers. This makes the jig especially useful at routing the edges of large stock because it's much easier to move the cutter past a piece rather than the other way around. I can also position a template quickly and accurately relative to the stock, which eliminates much of the trial and error that's required to set up some joint makers.

In addition, the jig's templates are mounted above the router (see the photo on p. 84), which makes them easy to see and keeps them away from the dust. Finally, the templates interchange quickly, and their holders easily adapt to different joinery, such as mortises and tenons, finger joints and dovetails.

Constructing the jig

Although it may look complicated, my joint-making jig was fairly easy and inexpensive to build (around $160, depending on the amount of work you have done by a machine shop). Basically, the jig is a plunge router mounted horizontally in an upright, linear-motion (X-Y) carriage, which is secured to a frame and table. A following device copies patterns secured by a template holder. I simply clamp the stock to the table, and trace the template with the follower as the router cuts the joint.

As shown in the drawing at right, the jig has five subassemblies: an X-Y carriage, a wooden frame that has a platform and a table, a horizontally adjustable template holder, a vertically adjustable follower, and a fence and hold-down to position and clamp stock. I sized the frame to suit my joinery needs, and then I built the rest of the jig around this.

To construct the carriage, you could buy the aluminum bar and flat stock from a metal supplier, but I picked up scrap aluminum for under 70 cents per pound. I cut

Router jig assembly _____

The router travels via linear-motion guides. A stylus traces the pattern while the router's plunge mechanism controls depth of cut. With workpiece clamped to table, operator stands on platform and moves router with two hands.

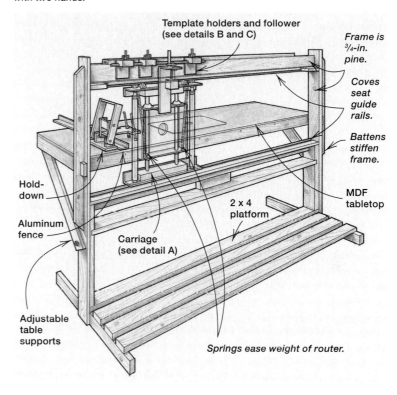

DETAIL A: X-Y CARRIAGE

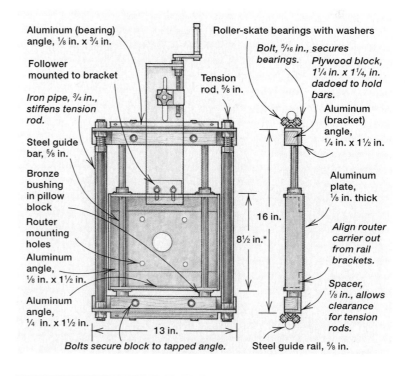

all the aluminum pieces to length on my tablesaw fitted with a carbide blade. Then knowing that I needed a few large holes in the pieces that I couldn't bore with my hand-held drill, I took the aluminum to a machine-shop equipped with a CNC mill/drill. The shop performed the work for only $30. Shops with conventional equipment gave me quotes around $100.

Carriage

The X-Y carriage consists of two major components: a vertical router carrier and a horizontal roller assembly. The router carrier holds the router and provides up-and-down movement by means of four bronze bushings that ride on two $5/8$-in.-dia. steel guide bars. Not expecting ever to have to rout more than 2-in.-thick tenons or dovetails, I allowed just $3^1/2$ in. of vertical travel. I mounted the bronze bushings in self-aligning pillow blocks made of stamped steel. The blocks are available from Northern Hydraulics, PO Box 1499, Burnsville, Minn. 55337; (800) 533-5545; or you could use linear-motion bearings, which are carried at most bearing-supply shops. The vertical bars are fastened to the horizontal roller assembly, which relies on four pairs of precision roller-skate bearings for motion. The bearings are bolted to $3/4$-in. aluminum-angle brackets. These bearing brackets are fastened to $1^1/2$-in. aluminum angles so that the bearings are oriented 45° on either side of two horizontal steel rails (see drawing detail A on p. 81). The bearings and rails work similarly to the guide system I used in a sliding saw table (see *FWW* #101, p. 51). I used 41-in.-long rails, which allow for 28 in. of horizontal travel.

The router carrier is made from four lengths of $1^1/2$-in. aluminum angle riveted together at the corners. I used $1/4$ in. angle for the upper and lower pieces of the carrier because they support the bronze bushings. A $1/8$-in.-thick aluminum plate mounted between the angles serves as a base for the router and stiffens the assembly. To keep the overall size down, I made the router carrier as small as possible, leaving just enough room for the router base to fit easily between the guide bars. Because the alignment of the bronze bushings is critical, I had their clearance and mounting holes professionally machined.

I connected the bearing-bracket assemblies with two $5/8$-in.-dia. by 18-in.-long threaded steel tension rods. The tension rods are stiffened by slightly shorter lengths of $3/4$-in. iron pipe. I had to cut some of the aluminum angle away so that the inside upper tension-rod nuts could turn freely. This allows just enough room to adjust the carrier for a tight fit to the guide rails. I secured the vertical guide bars to the horizontal brackets with two $3/4$-in. plywood mounting plates. I cut the bar-aligning dadoes from a single piece and ripped the two mounting plates from it.

Frame and table

After assembling and mounting the horizontal bearing brackets to the router carrier, I set the tension rods to allow for $1/2$ in. of adjustment either way. Holding the guide rails in place, I measured the outside distance to determine the inside height of the frame. I subtracted $1/4$ in. from the rail's out-to-out dimension to allow for cove cuts in the frame for the rails. I located the coves so the front of the carrier rides proud of the frame to provide clearance for machining longer stock. The frame width is determined by the length of the horizontal guide rails.

I initially built the frame from $3/4$-in. pine and later added battens to stiffen the frame. I think a 6/4 hardwood frame would be better. Also, I soon discovered that the frame provides a ledge for chips to build up on, so if I were to build the jig again, I would turn the frame boards on edge and cut bevels on either side of the lower guide rail.

The table is made of pine with a medium-density fiberboard (MDF) top, which can be slid away from the carriage to allow clearance. I also cut holes and slots in the table, so I could mount an aluminum fence and a shopmade hold-down (see the sidebar on the facing page). In the extra table space, I made a cutout for a vertical router. The lower braces support the table and keep it square to the carriage.

Template holder and follower

When I designed my machine, I was concerned with providing a way to hold the template and allowing crude lateral adjustments. And I knew that the follower should

A shopmade hold-down

I originally used Jorgensen adjustable bar clamps to hold down workpieces on my joint-making jig. But I soon found the repeated tightening and loosening of the clamps to be time-consuming and a real blister maker. I dismissed the idea of using toggle clamps because of their small size. Instead, I constructed my own clamp using scraps of hard maple, a piece of medium-density fiberboard (MDF) for a base, ⅝-in. threaded rod, dowels, screws and an assortment of ⁵⁄₁₆-in. bolts.

Building a hold-down is straightforward once you under-stand the basic operating prin-ciple. In the vertical clamp shown in the drawing below, the handle provides leverage to the clamping arm by means of a pivoting bracket, which is fixed between the arm and handle. The clamp locks in place when the handle's pivot point is pulled forward of the arm's pivot. But because clamping pressure diminishes as the arm pivot travels past the initial locking point, a travel-limiting stop is needed. The trick is in placing the stop so that the clamp locks down and exerts sufficient pres-sure. I arrived at a good balance (favoring clamping strength) by positioning and paring the block (crossbar) until I was satisfied with the locking action.

Because the forces in the hold-down are mostly vertical, I oriented the grain of the base bracket up and down to prevent splitting. However, because the bracket is screwed to the base, it's possible that the drywall screw could pull out of the bracket's end grain. To counter-act this, I reinforced the base brackets with hardwood dow-els. Holes drilled through the base enable me to bolt the hold-down to my jig's table. A pair of adjustable spindles with clamping pads resist any side-to-side movement of the workpiece. The spindles are two lengths of ⅝-in. threaded rod with top and bottom nuts. Rubber cap protectors (avail-able at most hardware stores) serve as the pads.

I use my oversized toggle clamp almost exclusively as a hold-down for my router jig, but I sometimes use it as a helping hand when I am power-sanding or freehand-routing. The clamp exerts a lot of down pressure, and I can quickly reposition the stock. The greatest virtue of the clamp, however, is its sheer size. Its long reach and big handle make the clamp truly a pleasure to use in repetitive operations. —G.P.

Shopmade hold-down

All parts are ¾-in. maple, except base, which is ¾-in. MDF.

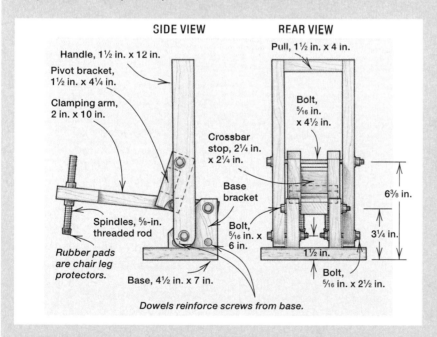

SIDE VIEW

Handle, 1½ in. x 12 in.

Pivot bracket, 1½ in. x 4¼ in.

Clamping arm, 2 in. x 10 in.

Spindles, ⅝-in. threaded rod

Rubber pads are chair leg protectors.

Base, 4½ in. x 7 in.

Crossbar stop, 2¼ in. x 2¼ in.

Base bracket

Bolt, ⁵⁄₁₆ in. x 6 in.

Dowels reinforce screws from base.

REAR VIEW

Pull, 1½ in. x 4 in.

Bolt, ⁵⁄₁₆ in. x 4½ in.

6⅜ in.

3¼ in.

1½ in.

Bolt, ⁵⁄₁₆ in. x 2½ in.

Router jig assembly details ———

DETAIL B: TEMPLATE HOLDER

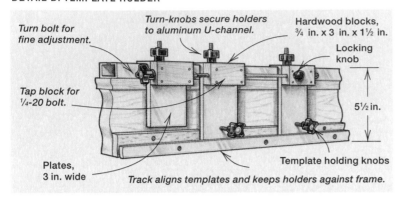

Turn bolt for fine adjustment.

Turn-knobs secure holders to aluminum U-channel.

Hardwood blocks, ¾ in. x 3 in. x 1½ in.

Locking knob

Tap block for ¼-20 bolt.

5½ in.

Plates, 3 in. wide

Template holding knobs

Track aligns templates and keeps holders against frame.

DETAIL C: TEMPLATE FOLLOWER

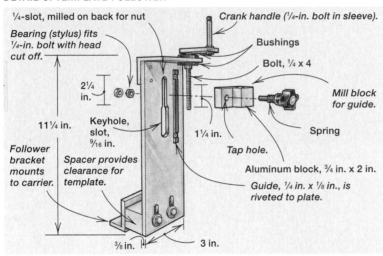

¼-slot, milled on back for nut

Bearing (stylus) fits ¼-in. bolt with head cut off.

Crank handle (¼-in. bolt in sleeve).

Bushings

Bolt, ¼ x 4

2¼ in.

11¼ in.

Keyhole, slot, 9/16 in.

Mill block for guide.

Spring

1¼ in.

Follower bracket mounts to carrier.

Spacer provides clearance for template.

Tap hole.

Aluminum block, ¾ in. x 2 in.

Guide, ¼ in. x ⅛ in., is riveted to plate.

⅜ in.

3 in.

be rigid and height-adjustable. For setup, I initially relied on a cut and nudge method: Take a trial cut, estimate the error and nudge either the template or follower to compensate. But it didn't take long to produce a pile of waste-tenons that way.

To remedy this situation, I introduced screw-driven adjusting mechanisms into both the template holder and the follower (see drawing details B and C at left). The template holders consist of three brackets constructed from ¼-in. aluminum plate and 1½-in. angle. The brackets can be individually locked to a guide track by turn-knobs (available from The Woodworkers' Store, 21801 Industrial Blvd., Rogers, Minn. 55374-9514, 612-428-2199).

The two right brackets are joined by a rod that adjusts for different-sized templates, and both are tapped for ¼-20 screws to mount the templates. The left bracket is fitted with a free-turning bolt that connects it with the template holders. Locking the left bracket only and turning the adjustment bolt moves the template 0.05 in. per turn. Recently, I got my hands on some FastTrack aluminum extrusions (available from Garrett Wade Co., 161 Avenue of the Americas, New York, N.Y. 10013; 800-221-2942), which when combined with their micro-adjuster and two micro-blocks made a nearly ready-to-use template holder.

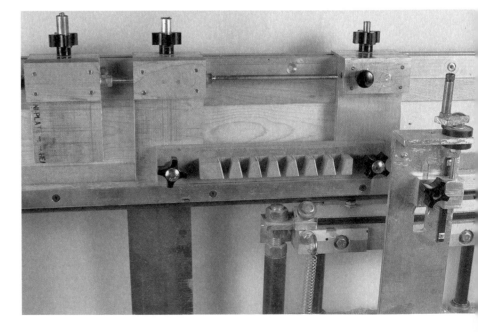

A pattern to follow— With a template bolted in place, the jig is ready to rout dovetail pins. The template holders (top) are adjustable left and right. The template follower (right) is height adjustable.

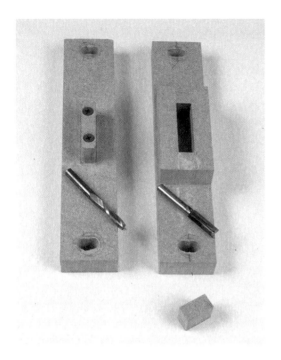

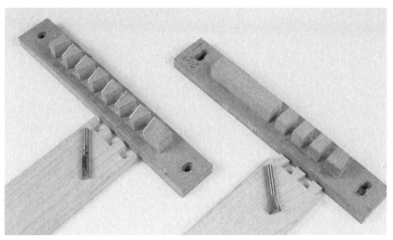

Matching templates and bits—As a sample of his jig's versatility, the author displays mated templates and corresponding bits. For adjustment, the mortise-and-tenon pattern (left) has a screw-on tenon and a mortise-shortening insert. The pin-and-dovetail templates produced the joinery examples above. When indexed by the tail pattern, the jig can cut finger joints; simply swap a straight router bit for the dovetail bit.

For the template follower, I had to add a means of inserting the bearing into the mortise template, so I devised a keyhole-shaped slot (see the photo on the facing page). After the follower bearing is slid into the narrow slot, it can then be cranked reliably, up or down, into position.

Routing mortises and tenons

Unlike the Leigh dovetail router jig, which uses adjustable templates, my jig has inter-changeable templates. I make my templates from scraps of medium-density fiberboard (see the top photo at left) because it is dimensionally stable, wears well and is easy to work. I make the mortise template first and then shape the tenon template to fit snugly into a test mortise. This is necessary because of the bearing system I use. Instead of ball bearings, I used a bronze bearing that slips on the $^1/_4$-in. follower shaft. I matched a $^1/_4$-in. mortising bit to a $^5/_{16}$-in. bronze bearing. This combination produces mortis-es that are slightly smaller than the tem-plate, so I fit the tenons to the mortise rather than to the template.

I usually eyeball the position of the tem-plate by first marking the stock, clamping it in place and then positioning the template

holder and follower so that the router bit just grazes my layout lines. Then I'll take a shallow test cut and measure the location with dial calipers. When I cut the stock, the surfaces that will be exposed are face down. I adjust the horizontal position by locking the left template bracket and turning the adjustment screw to move the template. I measure with my dial calipers to compensate for exactly half of the initial error. A similar technique adjusts the follower vertically.

When cutting mortises, the end of the stock bears against the router carrier. The edge is clamped against the fence with the hold-down. An aluminum plate fastened to the upper frame and scribed with a vertical indicator line marks the center of the cut. To lay out my mortises, I mark their center and align them with the indicator.

To cut the tenons, I bolt on a tenon tem-plate and change over to a $^1/_2$-in. straight bit. I climb-cut the first pass, which virtually eliminates any tearout and provides a very clean shoulder line. I complete the cut by merely following the template until no more shavings are produced. The X-Y carriage isn't stiff enough to entirely resist deflection, so I have learned merely to follow the tem-plate rather than force the follower bearing against it. I test-fit each piece immediately after machining, and I correct a too-tight fit by exerting a little more force during the cut.

Routing dovetails and pins

My joint-making jig handles through-dovetails (see the photo below) as easily as it does mortises and tenons. But making a set of dovetail templates is a bit more involved. The main trick is getting the spacing of the pin template to exactly match that of the tails. The tail template is really just a spacing guide, resembling half of a finger joint. In addition to getting the spacing and cut angle right, the pins of the template must be left full enough to ensure a tightly fitting joint. Also, by making the template at least twice as wide as the thickness of the stock, you can adjust the fit of the joint simply by changing the vertical position of the router bit on the stock. Because the height is relative to the template, it's easy to adjust the vertical position of the follower.

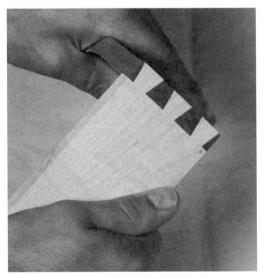

Machine-cut joint speaks for itself—Perez holds a drawer corner of oiled cherry and maple, dovetailed with his router jig. The jig cuts uniform dovetails or asymmetrical ones, if a hand-cut look is desired.

I've adapted Mark Duginske's method for cutting dovetail templates on the tablesaw (see *FWW* #96, p. 66). I use a set of wooden blocks to establish the spacing of the dado cuts. After cutting the tail template, I use the same set of blocks to machine the pin template. With this method, you can also make templates for non-uniformly spaced joints as long as you number and order both templates.

Always cut the tails first, using whatever dovetail bit the template is designed for (see the photo at right on p. 85). I place a piece of stock flat against the router carrier to set the depth of cut, extending the bit just proud of the piece, and clamp the stock face down. I adjust the template holders horizontally and position the cut so the outside tails are equidistant from the edges of the board.

I fit my router with a straight bit for machining the pins. I mount the template and position the stock so that the inside of the joint faces down. This arrangement ensures that once everything is adjusted, slight variations in stock thickness will not affect the joint's fit. The fit is determined by the distance between the follower and the bit—smaller distances will yield tighter joints and vice versa. With a test piece clamped in place (and the router unplugged), I position the follower so that the bit is just below the workpiece when the follower first contacts the bottom of the template. From here, I make test cuts and raise the follower.

PANEL ROUTER MAKES CUTTING DADOES A SNAP

by Skip Lauderbaugh

Many of my cabinetmaking projects require panels that have dadoes, rabbets and grooves to allow strong, easy assembly. I've tried lots of ways of cutting these joints and have found that a panel router is the quickest and most accurate tool to use. Unfortunately, the expense of one of the commercial machines (up to $3,500) and the floor space it requires (up to 25 sq. ft.) is more than I can justify. As is often the case, however, once you have tasted using the proper tool for a particular job, using anything else becomes a frustrating compromise.

I had seen other shopmade panel routers (for one example, see Steven Grever's article in *FWW* #88, p. 48), but they lacked features I wanted and seemed complicated. So I set out to design and build my own version of a panel router. By simplifying the guide system and by using common materials and hardware (see the drawing on p. 88), I built a panel router for less than $100 (not including the router, which I already owned). And although this jig easily handles big pieces of plywood and melamine, the jig folds compactly against the wall when it is not in use.

Wall-mounted panel router is ideal for making quick dadoes. Knowing his panel router had to save space, Skip Lauderbaugh mounted it to a wall at a comfortable height and angle. To build the jig, he used a router he owned and commercial hardware costing less than $100.

Panel-router assembly

Panel router handles common sheet thicknesses, stores flat against wall, folds out for use.

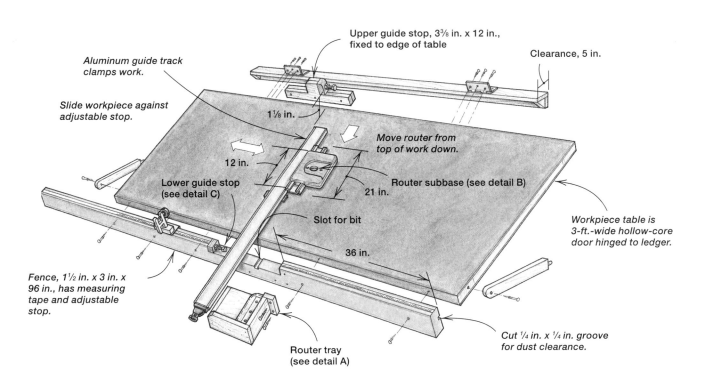

Upper guide stop, 3⅜ in. x 12 in., fixed to edge of table

Clearance, 5 in.

Aluminum guide track clamps work.

Slide workpiece against adjustable stop.

1⅛ in.

Move router from top of work down.

12 in.

Lower guide stop (see detail C)

Router subbase (see detail B)

21 in.

Slot for bit

36 in.

Workpiece table is 3-ft.-wide hollow-core door hinged to ledger.

Fence, 1½ in. x 3 in. x 96 in., has measuring tape and adjustable stop.

Cut ¼ in. x ¼ in. groove for dust clearance.

Router tray (see detail A)

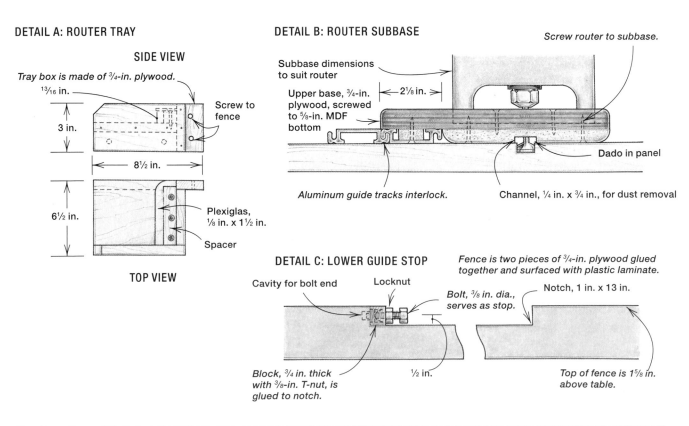

DETAIL A: ROUTER TRAY

SIDE VIEW

Tray box is made of ¾-in. plywood.

13/16 in.

3 in.

Screw to fence

8½ in.

6½ in.

Plexiglas, ⅛ in. x 1½ in.

Spacer

TOP VIEW

DETAIL B: ROUTER SUBBASE

Screw router to subbase.

Subbase dimensions to suit router

2⅛ in.

Upper base, ¾-in. plywood, screwed to ⅝-in. MDF bottom

Dado in panel

Aluminum guide tracks interlock.

Channel, ¼ in. x ¾ in., for dust removal

DETAIL C: LOWER GUIDE STOP

Fence is two pieces of ¾-in. plywood glued together and surfaced with plastic laminate.

Cavity for bolt end

Locknut

Bolt, ⅜ in. dia., serves as stop.

Notch, 1 in. x 13 in.

Block, ¾ in. thick with ⅜-in. T-nut, is glued to notch.

½ in.

Top of fence is 1⅝ in. above table.

Designing the panel router

Because the guide rails used in industrial panel routers often get in the way, the rails were the first things I eliminated on my design. The next thing was to orient the machine so that gravity would help feed the router into the work. Big panel routers are oriented horizontally, and they have the capacity to handle 36-in.-wide pieces of plywood. But because shelf dadoes in cabinets and cases are usually less than 3 ft. wide, I scaled things down a bit, and I situated the whole setup vertically. This orientation also saved considerable shop space. Then I came up with a clamp-on router guidance system, so I don't have to do any measuring or marking on a panel. Finally, I devised a router subbase that eliminates depth-of-cut adjustments when changing material thicknesses. To help you understand the abilities of this tool and how it is constructed, I've divided it into six basic components:

1. The workpiece table
2. The router guide system
3. The fence with adjustable stop
4. The upper and lower guide stops
5. The router subbase
6. The router tray

The workpiece table

A panel router requires a flat, stable work surface with a straight edge for mounting the fence. I chose an ordinary 3-ft.-wide hollow core door for the table because it provides those things, and at $15, it cost less than what I could build it for. I mounted the table to a ledger on the wall. The ledger is 75 in. from the floor to give a comfortable working height. A 5-in. space from the wall gives enough clearance for the guide system. Standard door hinges let the table swing out of the way during storage, and side supports hold the table at a 65° angle when the table is in use.

The router guide system

Several years ago, I discovered that the aluminum extrusions used in Tru-Grip's Clamp 'N Tool Guides (manufactured by Griset Industries Inc.; see Sources of Supply below) interlock when one is inverted (see the photo below). In this configuration, the two pieces slide smoothly back and forth with little side play, like a track. This system has several benefits: A panel can be set directly on the table without having to go under fixed guide rails. The guide is accurately located, and the panel is clamped tightly to the fence and to the table. The clamps are available in several lengths, but I've found that 36 in. is the most convenient (see the sources box). The manufacturer recommends using silicone spray to minimize wear.

The key to the router guide is interlocking aluminum track. When the author discovered the edges of Clamp 'N Tool Guides nest and slide easily, he made them into a two-piece guide system: An inverted 21-in. piece is fixed to the router subbase, and another piece is clamped to the work.

SOURCES OF SUPPLY

CLAMP 'N TOOL GUIDE
Griset Industries, Inc., PO Box 10114, Santa Ana, CA 92711; (800) 662-2892

ADJUSTABLE STOP
Biesemeyer, 216 S. Alma School Rd., Suite 3, Mesa, AZ 85210; (800) 782-1831

PANEL-ROUTER BITS
Safranek Enterprises, Inc., 4005 El Camino Real, Atascadero, CA 93442; (805) 466-1563

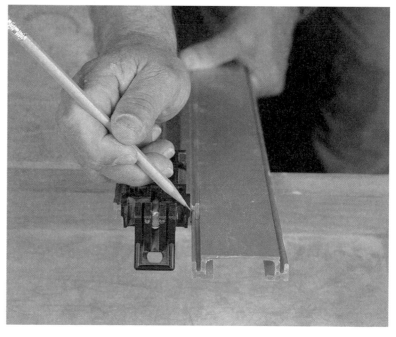

The fence's adjustable stop ensures perfect alignment. A Biesemeyer micro-adjustable stop and measuring system precisely positions the left side of the work for each dado or groove. Lauderbaugh uses a pair of dividers to point out two cursors that indicate left and right limits of a cut.

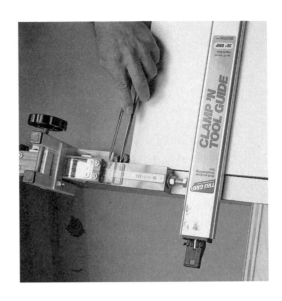

Channels align sub-base and evacuate dust—The underside of the router sub-base reveals an inverted aluminum guide channel and a medium-density fiberboard bottom with dust-evacuation slots cut across it for the bit.

The fence with adjustable stop

The fence holds the bottom edge of a panel straight, adds a runner for an adjustable stop and measuring system, and gives a place to mount the lower guide stop. Fence construction is partially dictated by the stop you use. I chose a Biesemeyer miter stop because it has two adjustable hairline pointers, which let you set and read both sides of a dado (see the photo at right above).

For the adjustable stop to work, the fence should be $1\frac{1}{2}$ in. thick and the top edge of the fence has to be $1\frac{5}{8}$ in. above the top of the table. My fence is two thicknesses of $\frac{3}{4}$-in. plywood laminated to form a $1\frac{1}{2}$-in.-thick piece that is 3 in. wide and 96 in. long. To allow the router to pass through at the end of a cut, I made a 1-in.-deep notch in the fence. The notch is 13 in. long to fit my router. I located this notch 36 in. from the right, so I can dado in the center of an 8-ft.-long panel. To finish off the fence, I glued plastic laminate to the top, faces and ends. Before mounting, I cut a $\frac{1}{4}$-in. by $\frac{1}{4}$-in. groove in the back to provide for dust clearance, which ensures that the bottom of a panel stays flush to the fence. The fence is mounted to the bottom edge of the table with $2\frac{1}{2}$-in.-long screws.

The upper and lower guide stops

The upper and lower guide stops allow the Clamp 'N Tool Guide to be set exactly 90° to the bottom edge of a panel. The lower guide stop is integrated in the fence, and the upper guide stop is fixed to the top of the table. The lower stop is a $\frac{3}{8}$-in. bolt threaded into a T-nut inset into a block and glued to a notch in the fence. The center of the bolt head should be $1\frac{1}{8}$ in. above the work surface, or $\frac{1}{2}$ in. above the bottom of the notch. The upper stop consists of two pieces of $\frac{3}{4}$-in.-thick plywood laminated to form a

$1^1/_2$-in.-thick piece, 12 in. long. The top is notched on both ends to leave a 2-in.- by $2^1/_2$-in.-wide section in the center. Another bolt and T-nut are screwed to the shoulder. The center of this bolt is $1^1/_8$ in. above the bottom of the notch. To fine-tune the stops for square, turn the bolts, and lock them with a nut. After the stops are set, adhere the measuring tape for the adjustable stops onto the top of the fence.

The router subbase

Parts for the router subbase consist of a medium-density fiberboard (MDF) bottom, an upper base made out of $^3/_4$-in. plywood that mounts to the router, and a piece of upside-down extrusion screwed to the side so it can engage the guide track. Drawing detail B shows the dimensions I used to mount my Porter-Cable model 690 router. But you could modify the subbase to suit your router. Regardless of the router, the bottom should be $^5/_8$ in. thick so that the extrusions interlock properly.

After the bottom is cut to size, center the baseplate on the bottom, and align the router handles at a right angle to the extrusion. Drill and countersink the mounting holes and mount the upper base to the bottom. Next, carefully, plunge a $^3/_4$-in. bit by slowly lowering the router motor. Then cut two dadoes, each $^1/_4$ in. deep by $^3/_4$ in. wide across the bottom. The first dado runs the full length and the second goes halfway across, 90° to the first. This T-shaped slot removes dust from the subbase (see the photo at left on the facing page).

For the piece of inverted extrusion, I obtained stock from the manufacturer. But because they currently don't sell this separately, just buy a 24-in. clamp, and cut off the ends. I used a 21-in.-long piece.

The bottom of the router subbase slides directly on the face of the panel so that the depth of cut is registered from the top of the panel. This is desirable because when you switch material thickness from $^5/_8$ in. to $^3/_4$ in., for example, the depth of cut does not have to be adjusted. Also, if the panel is slightly warped or some dust gets between

Commercial bits make clean cuts

Commercial panel routers work so well because the router bits are specifically designed to eliminate chipping and tearout, and they can also cut at higher feed rates. But their biggest benefit is that their cutter and arbor are two separate pieces (see the photo below), which means that the arbor can stay secured

Panel-routing bits change easily. The only things the author uses from industrial panel routers are the bits, which have interchangeable cutter tips.

in the router collet while you simply unscrew the cutter from the ½-in. arbor to change the bit size. Commercial panel-router bits (see Sources of Supply on the p. 89) are available in a full range of sizes, including undersized ones for veneer plywood and oversized ones for two-sided melamine. An arbor and cutter set costs about $35, less than a decent-quality dado blade set.

When you need to change the width of a dado, select the correct cutter size, and screw it on the arbor (no wrenches required). The depth of cut doesn't need to be reset because the height of the cutter stays the same. This process is much quicker than using a dado blade on the tablesaw, where you have to use shims to get the proper width, and then make test cuts to set the depth of cut. —S.L.

the panel and the table, the cutting depth is not affected. Interchangeable bits also speed up the process (see the sidebar above).

The router tray

The purpose of the router tray is to give the router a place to rest after it has completed a cut. The tray is mounted to the fence on the back side of the notched-out area. My tray is made out of $^3/_4$-in. plywood and is screwed

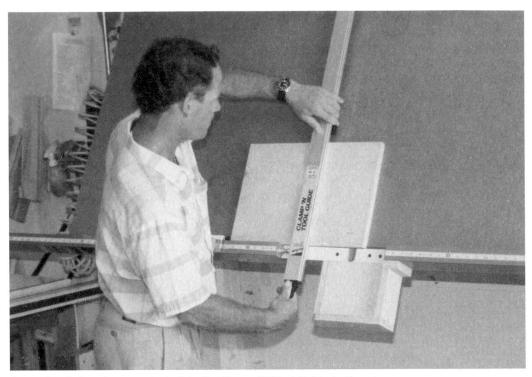

Setup for dadoes is easy. Just slide the Clamp 'N Tool Guide to the stops, and clamp the guide to the work by snugging up the black plastic dogs.

to the fence. On the right edge of the tray, a piece of $\frac{1}{8}$-in. Plexiglas protrudes into the tray opening. As the router slides down into the tray, the Plexiglas piece fits into a slot cut into the edge of the subbase and prevents the router from lifting out of the tray.

Using the panel router

The panel-router sequence to make a dado goes like this: First, I set the adjustable stop to locate the dado where I want it. Second, I set the panel on the table and slide it up against the adjustable stop. Third, I place the Clamp 'N Tool Guide on the panel, slide it against the upper and lower guide stops, and clamp it down (see the photo above). In this one step, the guide is squared to the panel and clamped to the table. Fourth, I set the router on the panel with the extrusions interlocked. I hold the router subbase above the top of the panel so the bit clears. Finally, I turn the router on and cut the dado. To make stop dadoes, I insert a spacer block in the bottom of the tray to prevent the router from cutting all the way across a panel. While this setup may not be perfect for a large production shop, it is certainly affordable and conserves space.

SHOPMADE TRIMMER FOR FLUSH-CUTTING EDGE-BANDS

by Jim Siulinski

Every time I use sheet goods to make cabinets, I'm faced with the job of banding the exposed edges. I'm turned off by glued veneer tape because I worry that it will eventually peel away or chip off. It also has that department-store furniture look. I prefer a solid wood edge-band, which is more durable and more attractive.

Carriage for trimming solid wood edge-bands

This carriage was designed to improve the stability of a trim router while cutting solid wood edge-bands flush with panels. The base rides on the face of the panel. The plane body and handles make a sure and comfortable grip. The fence guides the trimmer along the adjustments to the depth of the cut.

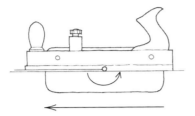

CUTTING DIRECTION
Use the trimmer with the bit turning into the cut to avoid tearout. Go slowly and steadily because the bit can self-feed.

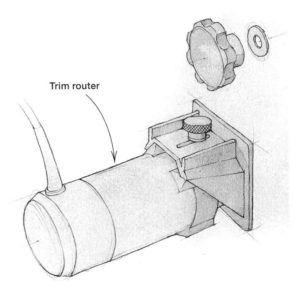

Trim router

ADJUSTING THE DEPTH OF CUT
The trimmer's performance depends on how evenly the bit cuts with the bottom of the base. It must be finely adjustable.

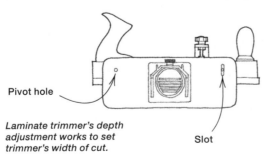

Slot in fence allows the base to travel up and down, pivoting on the opposite carriage bolt.

Pivot hole

Laminate trimmer's depth adjustment works to set trimmer's width of cut.

Slot

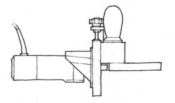

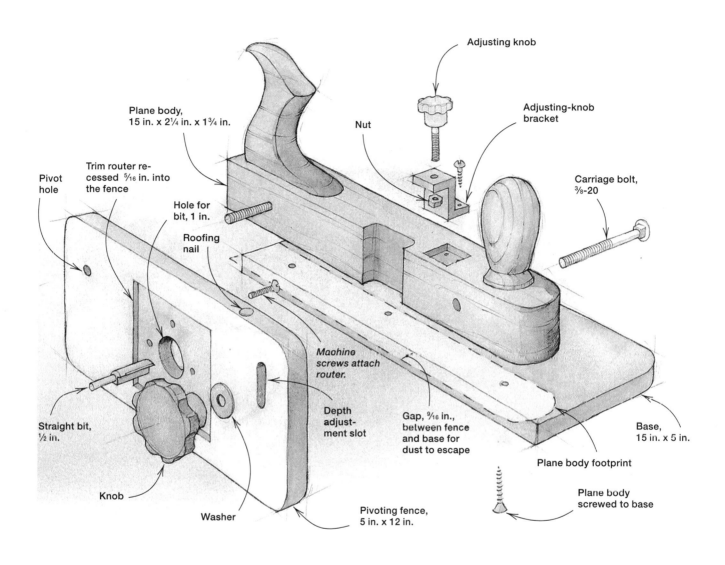

Plane body,
15 in. x 2¼ in. x 1¾ in.

Adjusting knob

Nut

Adjusting-knob
bracket

Pivot
hole

Trim router re-
cessed ⁵⁄₁₆ in. into
the fence

Hole for
bit, 1 in.

Carriage bolt,
³⁄₈-20

Roofing
nail

Machine
screws attach
router.

Straight bit,
½ in.

Depth
adjust-
ment slot

Gap, ⁹⁄₁₆ in.,
between fence
and base for
dust to escape

Base,
15 in. x 5 in.

Knob

Washer

Pivoting fence,
5 in. x 12 in.

Plane body footprint

Plane body
screwed to base

A simple tool for cleaning up banded plywood edges—The author devised a carriage that improves stability for his trim router and makes flush cuts a breeze.

Applying and trimming solid-wood edge-banding, though, can be difficult and time-consuming. After applying an oversized strip to the edge, you have to trim it flush with the face of the panel. I find it difficult to balance a router or laminate trimmer on a panel edge, and I immensely dislike sanding out the inevitable snipe and chatter marks from router wobble and bearing hops. I looked for a way to improve the process.

The solution is a stable carriage for the trim router

My solution was to make a carriage for a trim router with an extended base and fence and handles like those on a handplane (see the drawings on pp. 94-95). The trim router is mounted in the fence and attached to the base at 90°. The base rides on the face of the panel, and the fence rides along the edge. The 15-in. by 5-in. base significantly increases the surface area of the tool. It's stable and wobble-free. An adjusting knob (see the photo at right below) set into the top of the plane body allows precise alignment of

the trimming bit with the bottom of the base for a perfectly flush cut.

I scrounged most of the materials from a junk pile at my workplace and from a friend's woodshop. I used melamine with a medium-density fiberboard (MDF) core for the base and fence because it's stable, durable and slides well over the work. The wood in the plane body is jarrah, though any stable hardwood will do. The only uncommon part is a scrap of anodized-aluminum angle bar I used for a bracket to house the adjusting knob. If I hadn't found the angle bar, I probably would have made some kind of bracket out of wood. Like many shopmade jigs, this one is fast, easy and inexpensive to build. The whole jig took about four to five hours, start to finish.

It works much like a handplane

With the edge-banded plywood lying flat on a workbench, I use the carriage much like a handplane. To avoid tearout, the trimmer should be used with the bit turning into the cut, in the same direction as the trim-

Easy design and assembly from odd materials. Trimmer is mounted in a fence and attached to a base at 90°.

Adjusting knob sets cutting height. A roofing nail makes a resilient contact point, reducing wear on the fence.

mer's movement (see the drawings on pp. 94-95). This means that the carriage must be used in a left-handed fashion (lefties should appreciate this). Facing the work on a bench, start at the left, and move to the right. The mass of the carriage and the sure grip of the plane-like handles make it easy to keep the bit from self-feeding and clogging or skating down the workpiece. Be sure to clamp your work to the bench. A few test-cuts should ensure proper bit alignment with the base. I like leaving the band ever so slightly proud—just in case—and afterward, lightly sanding it flush.

Few or no obstacles to a clean cut

One of the trimmer's major advantages is its ability to trim directly over dadoes. I think it is easier to cut a dado prior to edge-banding, thus avoiding a more complex stopped dado cut. Using a trim router with just a bearing for a guide would ruin the edge as the bit turned into the dado.

Another advantage is that the trimmer is not thrown off by dried glue. A $^9/_{16}$-in. gap between the edge of the base and the fence makes it unlikely that any dried glue squeeze-out will interfere with the carriage base. A bearing-guided bit would create bumps in the edge-band as the bearing rolled over drips. It should be noted that a warp in the sheet will alter the trimmer's cutting depth, so clamp your work flat to the work surface.

The trimmer carriage works best when edge-banding sheet material at least as large as the carriage. I typically use it when making bookcases and shelves. Because the essential use of the fence is to make a stable cut, the carriage may be adapted to many other applications. I sometimes use it to trim the edge of a face frame on a finished case. By adjusting the bit, you can use it to cut rabbets. By changing bits, you can apply different molding profiles—and not just to edge-bands.

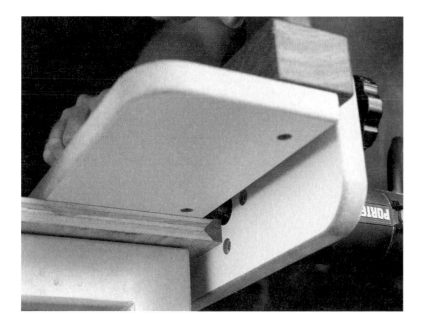

Gap between base and fence avoids obstacles. Trimmer won't hang up on glue squeeze-out or oversized edge trim.

FOUR

Jigs for Tablesaws

Tablesaws are almost invariably at the heart of small woodshops, although there are exceptions. Some chairmakers don't need them because they don't saw but split their wood. A few prefer a bandsaw for ripping and a chopsaw for crosscuts. But the otherwise universal appeal of the tablesaw makes these kinds of shops stand out. A shop without a tablesaw smack-dab in the middle looks naked, incomplete, or strangely specialized.

Tablesaws have such an appeal for the same reason routers do: they are adaptable to a wide range of cuts. The tablesaw's range of duties is somewhat more fundamental than the router's. The tablesaw can rip large stock down to size more efficiently than any other tool. At the same time, it's wrong to think of the tablesaw as a rough cutting machine. It can be tuned—and jigged—for very precise work.

This chapter can only touch on the wide range of possible jigs for the tablesaw. With various extensions to the table itself, the saw can accommodate larger pieces without straining your back and sacrificing accuracy. With sleds, boxes, and jigs that run in the miter gauge slots or register against the rip fence, the saw can accommodate smaller pieces, which otherwise might get caught between blade and fence. These jigs also make a wide range of joinery cuts possible, from tenons and finger joints to miters and corner slots. Perhaps most important, safety guards can be attached to these sleds and jigs.

Why make your own rip fence when aftermarket fences are accurate, if a little expensive? For the same reason that you'd make any jig in your own shop. You don't have to settle for an expensive compromise. A shop-made rip fence can be adapted to your particular way of working. And auxiliary jigs can be attached to a rip fence made of wood instead of steel.

MAKING A SLIDING SAW TABLE

by Guy Perez

Until I came upon a 9-in., used tablesaw (a 1937 Craftsman model), I cut all my wood with an 8¼-in. circular saw aligned by a pair of shopmade guides. But even the Sears tablesaw still lacked a stand, table extensions and a miter gauge. So I set out to bring my bargain saw up to a higher standard.

The first additions I made to the saw included a stand, table extensions and a T-square fence, which allows me to rip stock up to 32 in. wide. These improvements served me well through several furniture projects, but I continued to crosscut with my circular saw and guide instead of using a miter gauge. As I saw it, standard miter gauges have three weaknesses: Their bars often fit loosely in the miter slots, they don't support long pieces well and they're ineffective for crosscutting wide pieces,

A sliding table improves crosscutting and mitering. Guy Perez made this sliding table to extend the usefulness of his old Sears contractor's saw. Using lightweight everyday construction materials like plywood, pine and aluminum angle, Perez built the table with an adjustable fence, which makes the jig ideal for multiple crosscutting and for mitering.

especially sheet stock. I decided a sliding table would solve all of those problems.

But I found that most commercial sliding tables cost in excess of $350. The ones I looked at also failed to address another constraint I had—scarcity of shop space. So I built a scaled-down sliding table (see the photo on the facing page) that has a 32 in. crosscut capacity. The table cost me less than $100, but it performs comparably to the expensive commercial models. It was fairly easy to build, too, and I can still roll my entire tablesaw out of the way to save space.

How the table slides

Like a few of its store-bought cousins, my sliding table rolls on precision bearings that are guided by steel rails. The whole assembly (see the drawing below) consists of four main components: an 18-in. by 24-in. plywood table, an 18-in.-long carriage that has four pairs of bearings, two 5-ft.-long tubular guide rails and a 60-in.-long aluminum crosscutting fence. Similar to one of Robland's sliding tables, the rails are

spaced about 6 in. apart and are mounted left of the saw table. Drawing detail A on p. 102 shows the wooden frame I built to support the rails. You can easily modify the frame to suit the saw you have, as long as the sliding table is level with and travels parallel to the saw table.

Constructing the table

Before you start to build a sliding table, there are a couple of things worth noting about aluminum. First, when buying aluminum bar or angle, check out recycling centers and salvage yards because they usually don't require a minimum quantity or charge the premium that metal-supply shops often do. Second, you can cut aluminum to length on a tablesaw fitted with a carbide-tipped blade. But be sure you don't let the hot chips touch your skin.

Carriage

The carriage is the heart of the sliding table, so I built it first. It acts much like a sliding dovetail joint, in which a pin is held by the tapered dovetail groove, thus restrict-

Sliding table anatomy

The sliding table attaches to a carriage, which has bearings that ride along tubular guide rails (see carriage assembly). The rails are mounted to a frame that's fastened to the saw's stand (see stand detail). The pivoting fence (see fence detail) works for both crosscutting and mitering.

Recessed mounting bolt

To set miter cuts, pivot fence on left clamping block; slide right block in tabletop slot.

18 in.

Plywood top

Fence slides close to sawblade.

Turn knob on ¼-in. carriage bolt.

Stop block

24 in.

Bore ⁵⁄₁₆-in. hole for bearing bolt.

Fix stop block to spacer board.

Hex-head bolt, ⁵⁄₁₆ in. dia. by 2½ in. (with washer), slides in keyhole.

Fence, 60 in. long

14 in.

18 in.

48 in.

60 in.

Drill ¼-in. hole for bearing-bracket bolts.

Guide rails, 1-in. conduit, 6 in. on center

Spacer board, ¾ in.

Turn block serves as stop and as carriage release.

Sliding table details _____

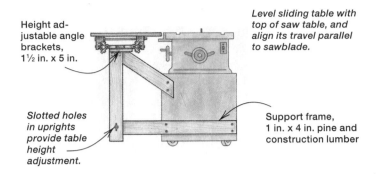

Height adjustable angle brackets, 1½ in. x 5 in.

Level sliding table with top of saw table, and align its travel parallel to sawblade.

Slotted holes in uprights provide table height adjustment.

Support frame, 1 in. x 4 in. pine and construction lumber

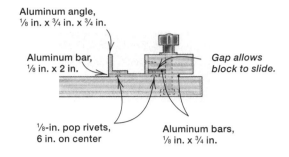

Aluminum angle, ⅛ in. x ¾ in. x ¾ in.

Aluminum bar, ⅛ in. x 2 in.

Gap allows block to slide.

⅛-in. pop rivets, 6 in. on center

Aluminum bars, ⅛ in. x ¾ in.

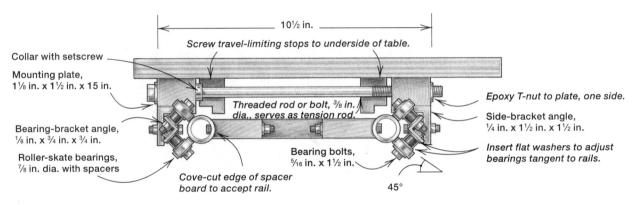

10½ in.

Screw travel-limiting stops to underside of table.

Collar with setscrew

Mounting plate, 1⅛ in. x 1½ in. x 15 in.

Threaded rod or bolt, ⅜ in. dia., serves as tension rod.

Bearing-bracket angle, ⅛ in. x ¾ in. x ¾ in.

Roller-skate bearings, ⅞ in. dia. with spacers

Cove-cut edge of spacer board to accept rail.

Bearing bolts, 5⁄16 in. x 1½ in.

45°

Epoxy T-nut to plate, one side.

Side-bracket angle, ¼ in. x 1½ in. x 1½ in.

Insert flat washers to adjust bearings tangent to rails.

---

ing side-to-side or up-and-down movement. In my sliding table, the pin is replaced by guide rails captured by opposing pairs of bearings (see drawing detail C above). By mounting the bearings 45° above and below the plane of the rails, I restricted both lateral and vertical carriage motion while allowing the table to roll forward and back.

I mounted two pairs of bearings and spacers to each of two bearing brackets (¾-in.-by 18-in.-long aluminum angles), putting one bearing and spacer pair at each end. The bearings and spacers I used are intended for skateboard wheels and are available at most sporting-goods stores and hobby shops for around $16 for eight bearings with spacers. I secured each bearing bracket to the vertical leg of the side brackets (1½-in.- by 14-in.-long aluminum angles) so that each leg of the bearing-bracket angle is 45° off the horizontal plane of the rails. I found it easiest to first drill the mounting holes in the

1½-in. side-bracket angles. Then, with the ¾-in. bearing-bracket angle clamped in place, I drilled its corresponding holes.

The side-to-bearing bracket assemblies are held together by two tension rods (I used hex bolts, but ⅜-in. threaded rod will also do). The tension rods allow the carriage to be precisely fit to the guide rails. The rods are fastened so that they turn freely in a mounting plate and are fixed to another plate with a T-nut that's epoxied in place. The 1⅛-in.-thick hardwood mounting plates are screwed to the top of each 1½-in. angle. Four T-nuts with bolts near the ends of the mounting plates secure the table to the carriage.

Guide-rail assembly

I built the guide-rail system to fit between the four pairs of carriage bearings. For the rails, I used 1-in. ID electrical conduit, which goes for about $4 per 10 ft. at building supply stores. I bolted a 6-in.-wide strip of $^3/_4$-in. particleboard as a spacer between the rails to stiffen them and keep them parallel. An easy way to determine the exact width of the spacer board is to set the tension rods, so there is $_1/_2$ in. of adjustment either way. Then hold the rails between the bearings, and measure the distance between the rails. Because the spacer board is cove cut on both edges to accept the tubing, add $^1/_4$ in. to the measurement. In each of the rails, I cut three keyhole-shaped slots by drilling sets of holes—each set with a $^1/_2$-in. hole overlapping a $^1/_4$-in. hole. The slots let me insert the hex bolts that fix the rails to the spacer board. I screwed a block at the rear of the spacer board to limit the table's travel. Then I added a turn block at the front, which lets me release the carriage from the guide rails.

The guide system is strong and accurate. Adapting a design from commercial-grade sliding tables, Perez made a roller-skate-bearing carriage and attached it to the underside of the table. The carriage is guided by a pair of tubular rails.

Support frame

To make the frame that supports the guide-rail system, I used eight board feet of 1x4 pine. I made the frame so I could easily get to all my saw's controls. Two 1$^1/_2$-in.- by 5-in.-long aluminum-angle brackets hold the rail assembly to the frame's uprights. Slotted holes in the uprights and oversized holes in the brackets provide the means for height adjustment.

Final adjustments

The carriage should be fit to the guide rails before you mount the sliding top. After adjusting the tension bolts so the bearings fit the rails, check two other things to ensure that the table will slide properly. First, the bearings should contact each guide rail tangentially. Second, each mounting plate should be parallel to the rail below it. I adjusted my bearings by using flat washers to get the proper spacing. To ensure parallel travel (vertically with respect to the rails), I used a combination square and set each mounting plate the same distance above the rails at the front and rear. Initially my carriage was misaligned. To fix this, I elongated the bearing-to-side-bracket holes using a rat-tail file, then I repositioned the brackets.

I made the final adjustments with the sliding table in place. To align the sliding table with the top of the saw table, I clamped 4-ft.-long straightedges across the front and

Crosscutting fence, extension and outfeed tables add to the accurancy, safety and versatility of the author's saw. Reduced friction and vibration of the sliding table are real assets when crosscutting long stock and sheet goods or mitering pieces—operations made possible by the added accessories.

rear of the saw table. I leveled the sliding table up to the straightedges and tightened the frame's height-adjustment bolts. Next I clamped a board to the sliding table, perpendicular to the rails. I drove a finishing nail in one end of the board, leaving it about $^1/_4$ in. proud. As I moved the table to and fro (with the saw unplugged), I measured from the nail head to the blade, both front and back. Once I was sure the table was parallel, I snugged up all the mounting bolts. Then I screwed travel-limiting stops to the underside of the table in line with the spacer-board blocks. To position the stop blocks, I rolled the carriage and marked limits for the table's normal movement.

Finally, I equipped my sliding table with a 60-in. crosscut fence (see drawing detail B on the facing page). Because the fence is adjustable, I can set it for mitering, and I can position it to support a workpiece right up to the blade.

SHOPMADE OUTFEED TABLE

by Frank A. Vucolo

Roller-topped drawers increase outfeed table capacity. By extending the bottoms of two drawers at the back of his tablesaw, Frank Vucolo created a place to mount outfeed rollers. Here, he opens one drawer to rip a piece of 6/4 mahogany.

In my small shop, ideal concepts are often compromised by the reality of limited space. My design for an outfeed table is a classic case in point. I started out thinking big. Ideally, I wanted the outfeed surface to extend 48 in. from the back of my tablesaw, so I would no longer have to set up and then reposition unstable roller stands. My ideal was quickly squashed, however, when I realized I couldn't dedicate that much permanent floor space. I need the space behind the saw to store my planer and router table when I'm not using them.

After some careful measuring, taking into consideration where I would locate all the machines, I concluded that the outfeed table should extend 30 in. from the back of the saw. But I still needed more support to rip long stock and to cut sheet goods.

While I was pondering possible solutions, I started to think about rollers that could extend off the back of the fixed table and then retract into it when they weren't needed. Then I remembered how amazed I was at the strength of Accuride's extension drawer slides (150-lb. capacity) when I had used them for file drawers in a desk pedestal. After a little more head scratching, nudged along by a couple of cups of coffee, I decided to incorporate the slides into a pair of drawers with rollers mounted on the front of them for the outfeed table (see the photo at left). Now I simply open a drawer to get an additional 24 in. of outfeed surface when I'm ripping long boards or cutting sheet stock.

Design and materials

Allowing an extra 1 in. for the extension rollers and the drawer slide action, the outfeed table is designed to support work up to 55 in. from the back of the saw table. With the drawers in the closed position, only 30 in. of floor space behind the tablesaw is committed. I made the drawers different widths so that I have various outfeed options, and I extended the drawer bottoms out in front of the drawers. This way, I have a place to mount the rollers (see the detail on the facing page). As a bonus, I get two drawers for storing saw accessories. And because the rollers are an integral part of the outfeed table, they are adjusted precisely in relation to the tabletop.

I constructed the outfeed table's top, legs and drawer bottoms out of ³/₄-in. birch ply-

Outfeed table assembly

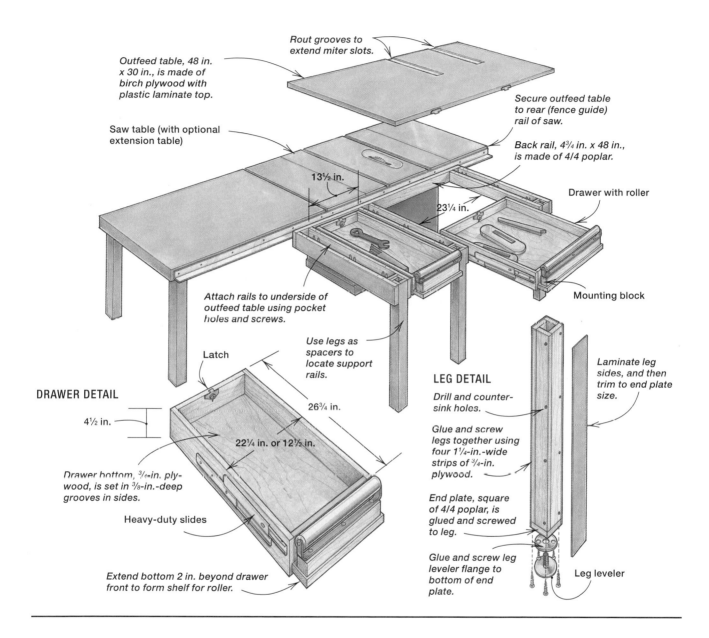

Rout grooves to extend miter slots.

Outfeed table, 48 in. x 30 in., is made of birch plywood with plastic laminate top.

Saw table (with optional extension table)

Secure outfeed table to rear (fence guide) rail of saw.

Back rail, 4³/₄ in. x 48 in., is made of 4/4 poplar.

13½ in.

23¼ in.

Drawer with roller

Attach rails to underside of outfeed table using pocket holes and screws.

Use legs as spacers to locate support rails.

Mounting block

Latch

DRAWER DETAIL

4½ in.

26¾ in.

22¼ in. or 12½ in.

Drawer bottom, ³/₄-in. plywood, is set in ³/₈-in.-deep grooves in sides.

Heavy-duty slides

Extend bottom 2 in. beyond drawer front to form shelf for roller.

LEG DETAIL

Drill and countersink holes.

Glue and screw legs together using four 1¼-in.-wide strips of ³/₄-in. plywood.

End plate, square of 4/4 poplar, is glued and screwed to leg.

Glue and screw leg leveler flange to bottom of end plate.

Laminate leg sides, and then trim to end plate size.

Leg leveler

---

wood. The under-table support rails are made from 4/4 poplar, as are the drawer sides, fronts and backs. For added protection and to give a nice slick surface, I covered the legs and top with plastic laminate.

To complete the material requirements, I bought the following hardware: two metal rollers, one 13 in. long and one 22 in. long (Wilke Machinery Co., 3230 Susquehanna Trail, York, Pa. 17402; 800-235-2100), two sets of heavy-duty drawer slides (I picked up Accuride's file-cabinet model from The Woodworkers' Store, 21801 Industrial Blvd., Rogers, Minn. 55374; 800-279-4441), three leg levelers (available from Woodworker's Supply Inc., 1108 N. Glenn Rd., Casper, Wyo. 82601; 800-645-9292) and a couple of latches (window sash locks), which I bought at a local hardware store. When you're determining the size of your drawers, keep in mind that the slides come in 2-in. increments, 12 in. to 28 in. long.

Drawer slide alignment is important. With the outfeed table flipped, the author positions a slide before he screws it to the poplar rail. Precise alignment ensures smooth operations of the outfeed rollers. A leg socket is below the square.

Level the outfeed table to match the saw table—After Vucolo secured the outfeed table to the rear guide rail of his saw, he turns the leg levelers (screw feet) to line up the two surfaces.

Making and mounting the table

To build the outfeed table, first determine the overall size (mine is 48 in. by 30 in.), and then cut the tabletop out of plywood. Temporarily mount the plywood to your saw, and level it using braces. This is so you can determine the length of the three legs. Measure each leg separately, and allow some room ($^1/_2$ in. or so) for height adjustment. The leg levelers will take up the play. Disassemble the table, and then fabricate the legs, as shown in the drawing detail on p. 105, including the plastic laminate.

With all three legs complete, lay out the support rail locations on the underside of the plywood top. Approximate the two different widths of the drawers plus their slides. Rip and crosscut the poplar pieces to size, and begin fixing the members to the plywood. I drilled pocket holes and then glued and screwed the rails in place. Start at one end, then use an assembled leg as a spacer to set the second rail. Next do the other end of the table, using another leg as a spacer. Set the two center rails in a similar fashion. Then attach the rear rail across the ends of the support rails. Also, cut and attach blocks behind each leg using the leg as a guide.

Mount the carcase portion of each drawer slide to the rails (see the top photo at left). Make sure you position all the slides the same distance from the bottom of the table. I used the rails as a reference. The drawers must be perfectly parallel to the top. While you have the table flipped, laminate the sides of the top, and trim them with a flush-trimming bit in a router. Turn the table over, so you can laminate and flush-trim the top.

Now temporarily mount the legs, and align the laminated table to your saw exactly as it will be positioned in use. Carefully mark the position of the miter slots on the top. Determine the depth of the grooves by referencing off the tablesaw. If you have a T-slot or dovetail-shaped miter-gauge runner, lay out the slots so that they will be a bit wider than the widest (bottom) part of the tablesaw slot. The outfeed table slots will be for clearance only.

Remove the outfeed table. Run the miter gauge all the way past the blade, so you can find the length of the runner as it hangs off the back of the saw table. Mark this length plus a bit extra onto the outfeed tabletop. If you use sliding jigs, like a crosscut box, check that their runners will work in the laid-out slot, too.

Using a straight bit and your router, cut the grooves in the surface of the outfeed table. A straightedge can be used to guide the router. But don't try to cut the whole depth in one pass. It's better to make two or three passes, removing a little at a time. Soften all the corners of the laminated top using a fine file. Also, ease the edges of the miter-gauge slots, and feather the edge that will go against the tablesaw. This will ensure that workpieces won't get hung up as they slide from the tablesaw onto the outfeed table.

How you mount the outfeed table to the saw will depend on the type of saw and fence guide rail you have. You can use angle brackets or drill directly into the rail. After you have the outfeed table in its approximate position, use a straightedge and a level to adjust the screw feet until the outfeed table is lined up to the saw table (see the bottom photo on the facing page).

Adding the drawers and extension rollers

The drawers should have a ³/₄-in. plywood drawer bottom extending 2 in. beyond the front of the drawer. This will provide enough rigidity for the extension rollers (see the drawing detail on p. 105). To receive the bottom, I plowed a ³/₈-in.-deep groove down the inside of each drawer side using a dado blade in my tablesaw. After I glued and screwed the bottom to each drawer, I butt-joined the front and back

Pocket holes and screws join drawer boxes—After temporarily clamping a drawer back, the author drives three screws into the sides using a flexible-shaft extension for his drill.

pieces together using pocket holes and screws (see the photo above). Then I attached the other part of the drawer slides to the outsides of the drawers.

It's critical that the rollers are mounted at the correct height. They should be at, or just barely above, the outfeed surface; they need to roll freely, without disrupting the travel of a workpiece. To get the proper height, I mounted the rollers using spacer blocks. First I set the roller on the shelf created by the extended drawer bottom. Then I measured from the top of the roller to the tabletop. I cut the block a bit oversized and then planed it down to thickness. If the roller is not parallel to the outfeed top and you can't adjust the drawer slides enough, taper the blocks slightly with the plane until the top of the rollers are level with the table. Finally, install a latch on the inside back of each drawer, so you can lock them in the open position.

ROLLER EXTENSION TABLE FOR HANDLING SHEET GOODS

by Bob Gabor

I was tired of wrestling big sheets of ply-wood across the top of my tablesaw. I already had an outfeed table on the back of the saw, but what I really needed was a side extension table to support the heavy panels going into the saw. I didn't want to give up too much valuable floor space to an accessory that I wouldn't be using most of the time.

My solution was a fold-away extension table that uses rows of roller balls to support the workpiece. I chose roller balls instead of long, tube rollers because the balls won't pull stock off-line as it is fed through the saw. Normally, the roller balls are even with the saw's tabletop, but they also can be raised to support long panels that overhang the end of my crosscut box. This straightforward shop fixture is easy to build and use. It sets up and drops back out of the way in a matter of seconds, and it makes cutting plywood on the tablesaw safer and more manageable.

Utility and economy in a shop tool

I'd rather make furniture than shop tools, so I designed the extension table to be as simple as possible. The top frame and the leg assemblies, as shown in the drawing on p. 110, are inexpensive and easy to assemble with a biscuit joiner. Yet they're light and strong. The length of the top-frame assembly and the leg assembly is determined by the distance between the floor and the top of the saw.

The top frame needs to be sized to just clear the floor in the folded position. The legs must be long enough to make the roller balls level with the saw top when the frame is in the raised position.

The extension table also supports long stock in my sliding crosscut box because the rollers are adjustable by the thickness of the crosscut box's bottom. Mounting the rollers on T-shaped assemblies, which adjust easily after loosening a few knobs, was a simple and reliable solution.

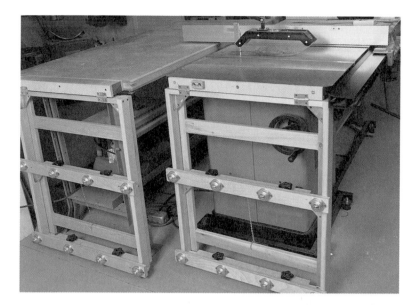

The extension table drops into its stored position in seconds and takes up no floor space. Adjustable roller assemblies can be raised so that the table also works with a sliding cross-cut box.

A roller-ball extension table makes cutting large panels safer and easier. Unlike long tube rollers, roller balls won't pull stock out of line as it goes through the saw.

Roller extension table

A shop-built extension table makes cutting large panels on the tablesaw a safer, easier operation.

ROLLER ASSEMBLIES

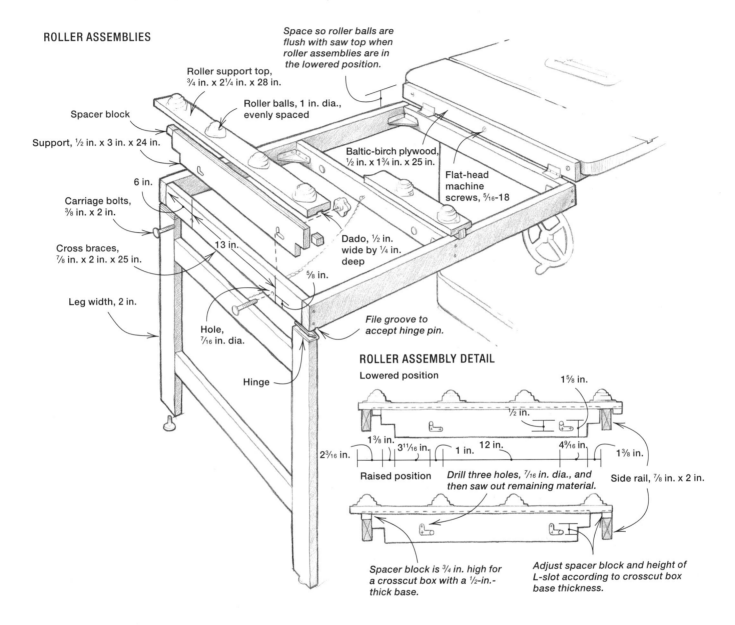

Space so roller balls are flush with saw top when roller assemblies are in the lowered position.

Roller support top, ¾ in. x 2¼ in. x 28 in.

Roller balls, 1 in. dia., evenly spaced

Spacer block

Support, ½ in. x 3 in. x 24 in.

Baltic-birch plywood, ½ in. x 1¾ in. x 25 in.

Flat-head machine screws, ⁵⁄₁₆-18

6 in.

Carriage bolts, ⅜ in. x 2 in.

Cross braces, ⅞ in. x 2 in. x 25 in.

13 in.

Dado, ½ in. wide by ¼ in. deep

⅝ in.

Leg width, 2 in.

Hole, ⁷⁄₁₆ in. dia.

Hinge

File groove to accept hinge pin.

ROLLER ASSEMBLY DETAIL

Lowered position

1⅝ in.

½ in.

1⅜ in.

2³⁄₁₆ in.

3¹¹⁄₁₆ in.

1 in.

12 in.

4⁹⁄₁₆ in.

1⅜ in.

Raised position

Drill three holes, ⁷⁄₁₆ in. dia., and then saw out remaining material.

Side rail, ⅞ in. x 2 in.

Spacer block is ¾ in. high for a crosscut box with a ½-in.-thick base.

Adjust spacer block and height of L-slot according to crosscut box base thickness.

To fold the unit for storage (see the top photo on p. 109), I hinged the legs to the top frame and also hinged the top frame to the tablesaw top. When folded down, the table doesn't take up much room in my shop. By adding adjustable levelers to the leg assemblies, I made it easy to fine-tune the height.

Finally, I added a piece of lightweight chain to limit the leg travel and a screen-door hook to keep the leg assembly folded for storage. I've been so pleased with the roller extension table that I've built another and attached it to the side of my out-feed table.

SIMPLE AND ACCURATE RIP FENCE

by Worth Barton

When I bought an older-model 10-in. Craftsman tablesaw, I was pleased with the saw's operation, but I was frustrated by its rip fence. It was a pitifully thin zinc die-casted saddle that soon broke. I could have bought one of the many after-market rip fences that are available (see *FWW* #133). But I knew that I could make a sturdy, accurate rip fence fairly inexpensively (mine cost about $65) following a few simple ideas (see the photo at right).

Design
Building a first-rate replacement rip fence is pure fun—good for the shop and for the ego. I began by making a "got to have" list:
- Strength, durability and deflection resistance
- Reasonably available components
- Construction requiring only a drill press and hand tools
- Repeatable settings with low-friction movement
- Consistent clamping behavior
- Quick removal

Using square tube steel for the fence took care of the first three items, and a toggle clamp in the saddle satisfied the last three.

Steel parts plus single-rail locking equal precision
Most impressive of the commercial rip fences are the cast-iron and steel ones that marry precise surfaces to smooth movement. I decided to capitalize on those principles using common steel sections bolted together. I also decided to use a single-rail

Rip fence slides smoothly, lock positively—Barton, setting up for a rip cut, snugs his fence in position using the saddle's toggle clamp. The fence, which has a hardwood auxiliary fence, was made out of standard steel sections and hardware. The fence slides on three brass setscrews that contact two square-tube guide rails.

locking mechanism. Here's why: Many commercial rip fences, once positioned, get locked to both the front and rear guide rails. But when I checked a couple of fences of this type using a dial indicator, I found that the locking action would slightly skew the fence out of parallel. So on my rip fence, I made a saddle that has a toggle-clamp plunger offset below two guide bolts at the front rail (see the drawing detail on p. 112). This provides the fence with the necessary

Rip fence construction

Note: Modify dimensions to suit your tablesaw.

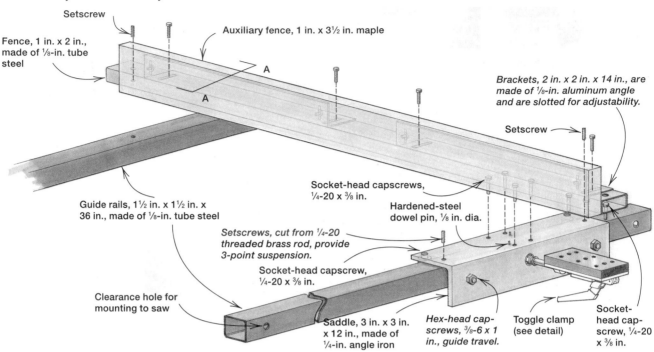

Setscrew

Fence, 1 in. x 2 in., made of ⅛-in. tube steel

Auxiliary fence, 1 in. x 3½ in. maple

A

A

Brackets, 2 in. x 2 in. x 14 in., are made of ⅛-in. aluminum angle and are slotted for adjustability.

Setscrew

Socket-head capscrews, ¼-20 x ⅜ in.

Hardened-steel dowel pin, ⅛ in. dia.

Guide rails, 1½ in. x 1½ in. x 36 in., made of ⅛-in. tube steel

Setscrews, cut from ¼-20 threaded brass rod, provide 3-point suspension.

Socket-head capscrew, ¼-20 x ⅜ in.

Clearance hole for mounting to saw

Saddle, 3 in. x 3 in. x 12 in., made of ¼-in. angle iron

Hex-head capscrews, ⅜-6 x 1 in., guide travel.

Toggle clamp (see detail)

Socket-head capscrew, ¼-20 x ⅜ in.

SECTION A-A

Optional hold-down bracket.

Tablesaw top

SADDLE AND CLAMP DETAIL

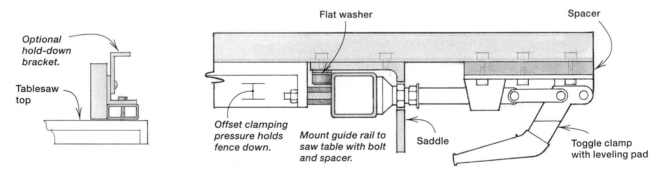

Flat washer

Spacer

Offset clamping pressure holds fence down.

Mount guide rail to saw table with bolt and spacer.

Saddle

Toggle clamp with leveling pad

down pressure (to the table), which means that I don't need a clip at the rear rail. Though the rip fence still requires a rear rail, it is for guidance only—not for latching the rip fence in position.

The saddle, the piece that connects the fence to the front guide rail, is really the key. I chose a long base for the saddle to control what aircraft and boat designers call pitch, roll and yaw. To picture these phenomena, think of the fence as an airplane's fuselage.

Pitch refers to the degree of nose-to-tail level. Roll is side-to-side (port to starboard) level. On the saw, these motions are relative to the table, the horizontal reference plane. To understand yaw, think of the fence's saddle as the tail of the airplane. The tail can swivel back and forth while remaining level with respect to the fuselage, as though the airplane was pivoting about a vertical axis. Yaw is similar on the saw, though it is greatly diminished; the rip fence can twist from side to side in the plane of the saw table, like a washing machine agitator.

To control yaw when the saddle position is fixed, I mounted two guide bolts (behind the front guide rail) and an adjustable De-Sta-Co toggle clamp between the two bolts (ahead of the front guide rail). The front rail is sandwiched between the ball-and-socket pad on the end of the clamp's plunger and the heads of the screws. That keeps the fence perpendicular to the rail. Because the screws in the saddle are above the plunger, clamping pressure forces the fence onto the saw table. This pressure enables the rip fence to resist the uplift action of a hold-down device. I use Shophelper hold-downs (available from Woodworker's Supply, 1108 N. Glenn Rd., Casper, Wyo. 82601), which have anti-kickback rollers. Two additional bolts, widely spaced and adjustable (see the photo at right), control yaw when I slide the saddle. This allows the motion to be smooth and free from lock-up.

Contact points: setscrews and guide rails

To restrict pitch and roll, I made the fence so it contacts the guide rails at three places: Two saddle points ride on the front rail, and one point on the end of the fence rides along the rear rail. For the contact points, I installed three brass setscrews, which provide a means of leveling and act as low-friction bearings.

The guide rails must be square or rectangular because the saddle has to lock positively to the front guide rail, and the setscrews must slide on flat surfaces at both the front and rear guide rails. If your tablesaw has pipe rails or angle-iron rails, replace them with square-tube sections (see the drawing on the facing page). To attach the rails, drill through the flange (edge) of your saw table, so you can use bolts and spacers to hold the rails perpendicular to the miter slots and parallel to the table surface. Set the rails lower than the table, so the rails don't interfere with the miter slots.

Materials, fasteners and assembly

Because steel has three times the stiffness of aluminum, I used standard structural steel box shapes for the long members (the two guide rails and the fence). The tube steel is dimensionally uniform, has a high resistance

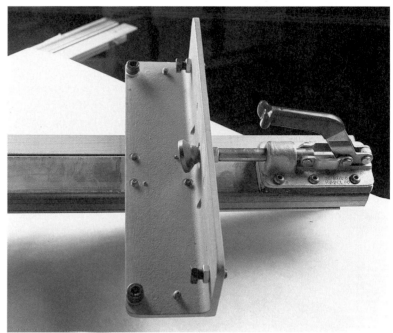

The underside of the saddle reveals how the fence stays aligned. Two socket-head screws (left side of angle) and a toggle clamp (right side of angle) sandwich the front guide rail. The saddle is fixed squarely to the fence by bolts and steel dowel pins. The two other bolts (with nuts) guide the fence during positioning.

to twisting and is readily available at most metal-supply houses (see Sources of Supply on p. 114). I used $^1/_8$-in.-thick wall tubing to avoid bolt tear-out in the tapped holes. Have your steel vendor cut the tubes to the exact length you need. For the saddle, I used a 12-in. length of $^1/_4$-in. by 3-in. by-3 in. angle iron. A model #607 De-Sta-Co toggle clamp locks the saddle; a piece of straight maple, attached by aluminum angles, serves as an auxiliary fence (see the drawing).

I bolted the parts together rather than welding them. Struggling with welding distortion can ruin your day. By contrast, bolts are easy to drill and tap for and are easy to remove. Suppliers offer a wondrous variety of fastening and clamping devices (see Sources of Supply). I use short fasteners because they reduce connection springiness and still afford some adjustability. Socket-head capscrews are ideal because they are made of high-quality steel and install easily.

The saddle-fence assembly is essentially a T-square, which glides on the front and rear rails. A 0.005-in. to 0.015-in. gap between the table and the fence allows clearance for sawdust and promotes smooth movement. As a safety feature, I extended the fence over the toggle clamp (see the photo on p. 111), which prevents me from accidentally bumping the actuating lever.

SOURCES OF SUPPLY

CLAMP

De-Sta-Co, PO Box 2800,
250 Park St., Troy, MI 48007;
(313) 589-2008.
(Note: modify thread to ⅜-16 for
stud-type plunger pad.)

ALUMINUM AND STEEL

Adjustable Clamp Co., 417
N. Ashland Ave. Chicago, IL
60622; (312) 666-2723

Castle Metals, 3400 N. Wolf Rd.,
Franklin Park, IL 60131;
(708) 455-7111

TOOLING ACCESSORIES AND FASTENERS

Reid Tool Supply Co., 2265 Black
Creek Rd., Muskegon, MI 49444-
2684; (800) 253-0421

Vlier Corp., 2333 Valley St.,
Burbank, CA 91505;
(818) 843-1922

Brackets secure the auxiliary fence and a hold-down accessory. The author used aluminum angles to attach the auxiliary fence to the main fence. The brackets are slotted to allow adjustment and to set minute tapers for riping. Similarly, the add-on top bracket is for mounting the yellow-wheeled (Shophelper) hold-downs.

Slotted brackets attach the auxiliary fence to the main fence (see the photo above). A similar bracket attaches the optional hold-down. The main brackets are symmetrical so that the auxiliary fence can be placed right or left of the fence. The slots, unlike holes, allow the fence to be adjustable and skewed for 2° or 3° tapers. It also enables the fence to be opened slightly at the rear of the saw (mine skews 0.020 in. over its length), as opposed to being parallel to the blade. A flared rip fence lessens the likelihood of kickback when you're ripping wet or unstable wood.

Setup for use

Make sure your sawblade and miter slots are parallel. Then set the fence parallel to the slots. To do this, place the assembly on the saw and attach a dial indicator to a miter-slot guide. Run the guide back and forth in the slot as you check the fence for runout. Tighten the screws joining the saddle to the fence. To ensure that the fence-saddle squareness won't be lost through rough handling, match-drill the parts so that you can press-fit hardened-steel dowel pins to lock the assembly: First clamp the saddle and fence together, and then drill and ream them to receive the pins. You can press in the pins with a drill press or tap them in with a hammer.

SLIDING FENCE FOR A MITER GAUGE

by Tim Hanson

The piece of scrap I kept bolted to my tablesaw miter gauge was a great improvement over the gauge alone, especially when making crosscuts. This extra fence made the gauge easier to grip, and it supported the workpiece right up to the blade. The problem came when making angled cuts. Each new angle made a new divot in the fence, and pretty soon, it looked like an old comb with missing teeth. I would try to save time by using one of the gaps as a point of reference when cutting, but sooner or later, I'd use the wrong one. Then I'd get ticked off and have to stop work to make a new fence, and the whole cycle would start again.

I finally took the time to make a fence that could be moved right or left and locked in place by simply flipping two little levers (see the photo at right). Now I can make minute adjustments in the position of a workpiece by releasing the levers and sliding the fence rather than unclamping and re-clamping. The fence makes using the tablesaw faster, safer and more accurate.

How it works

The wooden fence is held to the miter gauge by a pair of machine screws. The screws go through the miter gauge and are tapped into $2^1/2$-in.-long metal bars that ride in T-slots in the back of the fence. When the machine screws are loosened, the fence can be adjusted right or left—exactly where you want it. Flip the levers up, and the fence slides right up to the blade (see

The lever-action adjustment on this shop-built fence lets you position the fence quickly.

the photo on p. 116). Flip them down, and the fence is locked in place.

The fence is made from a clear, straight piece of 2x4 construction lumber. I made it 20 in. long thinking I'd shorten it later, but I found the length useful when crosscutting long pieces.

I used a tablesaw to make the T-slot, but there are other ways to do it. The important thing is to make the slot larger than the bar stock by about $1/16$ in. all around for easy sliding.

Fine-tuning the levers

I fashioned the levers from right-angle mending plates, which I purchased at the hardware store. The drawings on pp. 116-117 show the parts and how they go together, but the system does need some fine-tuning. Secure the levers under the heads

The sliding miter fence

Made from scrap lumber and easy-to-find hardware, this adjustable fence supports the work right up to the blade, no matter what the angle. It makes cross- and miter-cutting safer and more efficient.

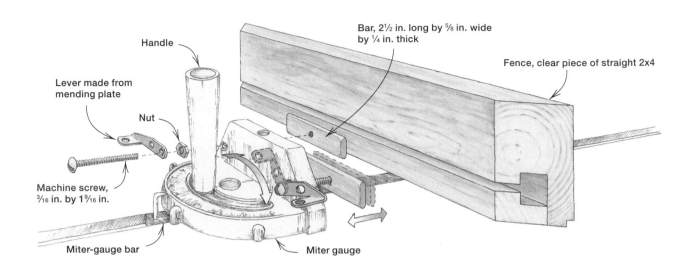

Handle

Lever made from mending plate

Nut

Machine screw, ³⁄₁₆ in. by 1⁹⁄₁₆ in.

Miter-gauge bar

Miter gauge

Bar, 2½ in. long by ⅝ in. wide by ¼ in. thick

Fence, clear piece of straight 2x4

This sliding fence is easy to grip with hands or clamps, and it supports the work right up to the blade.

of the machine screws with a wrench-tight nut. Slide the machine-screw assemblies through the miter gauge, and turn the bars onto the screws so the ends of the screws are flush with the bars.

Now turn one lever all the way to the left (at the 9 o'clock position), and slide the bar into the T-slot. Flip the lever to the right. The fence should tighten up against the miter gauge at about the 2 o'clock position without much effort. If it rotates past that point and the fence still isn't tight, the lever has to be repositioned. Disassemble the fence, and remove the machine screw. Clamp the machine screw between two blocks of wood in the vise, and loosen the nut just enough to rotate the lever counter-clockwise about one-quarter turn. Tighten the nut and reassemble. It may take a few tries to get the levers to grip and release in the correct position. Use the same procedures to adjust the second bar.

Precise cuts come from an accurate fence. My miter gauge's face wasn't perpendicular to the table, so I had to handplane the wooden sliding fence to make it square.

Sliding fence

Make a T-shaped slot in the fence. Cut the pieces to the dimensions shown at right and glue together. Make sure the bar stock moves freely in the slot. Miter the ends of the fence at 45°.

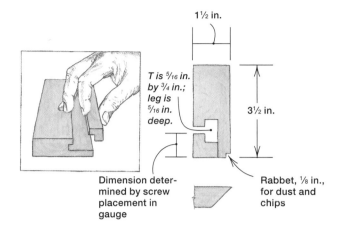

1½ in.

T is ⁵/₁₆ in. by ¾ in.; leg is ⁵/₁₆ in. deep.

3½ in.

Dimension determined by screw placement in gauge

Rabbet, ⅛ in., for dust and chips

Levers

Make two levers from 2-in. by 2-in. mending plates. Cut one leg about ½ in. long, and round all the corners. Make the left lever by bending the short leg toward the back of the visc, as shown. For the right lever, bend the short leg toward the front.

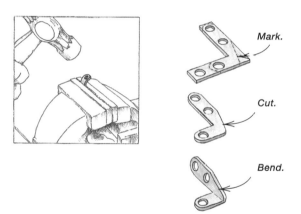

Mark.

Cut.

Bend.

Bars

Attaching screw to bar— A machine screw threads into the bar to lock the fence in position. If you don't have metal taps to make this connection, you can use a standard nut, as shown at right.

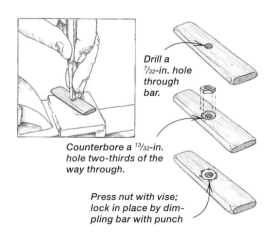

Drill a ⁷/₃₂-in. hole through bar.

Counterbore a ¹³/₃₂-in. hole two-thirds of the way through.

Press nut with vise; lock in place by dimpling bar with punch

SHOPMADE TABLESAW GUARDS

by Sandor Nagyszalanczy

"Blade guard removed for photo clarity." How many times have you been watching a home-improvement show or woodworking video and seen those words appear across the bottom of the television screen? Well, I want to know: *what* blade guard? In almost all the cases I've seen, a stock tablesaw guard wouldn't have worked in the applications shown.

What's a woodworker to do? Must we continually expose ourselves to unreasonable risks when we perform operations that require removal of the tablesaw's standard blade guard—jobs like sawing tenons, cut-

ting box joints and cove cutting? I suppose we can hope our luck holds out, or we can wait for some kind of sensational all-purpose saw guard to hit the market. But I advocate another alternative: to design safer tablesaw jigs and setups by adding guards and safety devices that prevent accidental contact with the sawblade. I think any woodworker bright enough to design innovative jigs for complicated woodworking tasks could make those same jigs a lot safer without investing too much extra time or material. After all, how much is a finger worth?

Safety without sacrifice—A Plexiglas shield keeps hands safely away from the blade without compromising visibility on the author's box-joint jig.

In this article, I'll show you some of my solutions for making common tablesaw jigs and setups much safer. One thing I aim for in modifying my jigs is to reduce the degree to which safety relies on judgment. It's a given that, as you work, especially at repetitive tasks, there will be times when your attention flags or is diverted. A safe jig protects you during these lapses. The very best safety feature is one that eliminates the possibility of contacting the blade with anything but the stock. I try to get as close as possible to this ideal in all my jigs.

In many cases, I've retrofitted existing jigs with guards to show that you don't have to build all new devices to add safety to your woodworking. Because jigs are, by definition, custom-made, the safety measures you take will also have to be individualized. So I haven't tried to cover all the bases here, only to share a few specific solutions and underscore the general idea that safety and guarding features ought to be built into every jig you make.

Clear guards for sliding jigs

Carriages that slide in the tablesaw's miter slots almost always require that the stock guard be removed. Whether you want to use a sliding crosscutting box or a jig for cutting tenons, dovetails or box joints, you can easily retrofit clear blade guards that allow you to see what's going on but keep you from getting cut.

Box-joint-jig guard

I made the guard for my box-joint jig shown in the photo on the facing page in about a half-hour from a few scraps of wood and a Plexiglas cutoff purchased from a local plastics store. (Glass shops and hardware stores often carry clear plastic sheet goods.) The guard is a low box with wood sides and a Plexiglas top that mounts directly over the box-joint jig and provides protection ahead of and after the cut. As an added bonus, I've noticed that it deflects chips and makes dust-collection more efficient.

I made the guard's frame 21 in. wide by 10 in. long, which is wide enough to handle 10-in. drawer sides. I drilled holes in the 1/8-in.-thick Plexiglas sheet so that it could be screwed to the top of the frame

Rear guard action— The simple outrigger behind the box-joint jig lets you complete the cut without exposing the blade.

(leave the protective paper on the Plexiglas during cutting and drilling to protect it from scratches). When attaching the plastic, I left it about an inch shy of the face of the jig, creating a slot for the workpiece. The 2-in.-high sides provide plenty of clearance between the plastic and the blade. I chamfered and waxed the lower edges of the sides to keep them gliding smoothly. Then I attached the guard to the back side of the box-joint jig with screws through the rear frame member.

To provide blade protection behind the jig, I added a second guard made from a block of wood and a 3-in. by 4-in. piece of Plexiglas, screwed to the underside of the rear frame member (see the photo above). Even if you don't want to make the entire guard frame, adding a rear guard is an excellent idea. It protects you after the jig has been pushed through the cut when you're reaching over the saw table and are probably the most vulnerable to blade contact.

This type of exit guard is a good addition to any sliding jig. And you can make using it even safer by clamping a stop block to the rip fence or right to the table that will limit the forward travel of the jig—allowing a complete cut through the workpiece but stopping the blade short of the exit guard's rear block.

Tenoning-jig guard

Protecting my hands from the blade involved the addition of three components to my sliding tenoning jig: a clear plastic

Untouchable tenoning jig—An adjustable Plexiglas blade guard and a hand rest combine to keep your exposure to the blade near zero on this tenoning jig. The block that's clamped to the rip fence provides a positive stop and prevents the blade from cutting through the exit block at the back of the jig.

shield ahead of the cut, an exit block to cover the blade behind the cut and a hand rest to prevent my left hand, which holds the workpiece against the jig, from sliding down into harm's way, as shown in the photo at left. The clear shield is nothing more than a 10-in.-long, $2\frac{1}{2}$-in.-wide piece of $\frac{1}{8}$-in.-thick Plexiglas screwed to the edge of a wood strip. This strip mounts to the face of the tenoning jig via slotted holes I made using a straight bit in the plunge router. The slotted holes allow me to shift the shield in or out depending on the width of the workpiece. I glued and screwed a $2\frac{1}{2}$x2x$1\frac{1}{2}$ wood exit block to the back of the jig. I used a brass screw just in case it's accidentally hit by one of the two sawblades used during tenoning. A larger block would provide more protection, but as long as you use the jig in conjunction with a stop block, this size is fine. The final component, the hand rest, is a 4x2x$1\frac{1}{2}$ block glued to the edge of the tenoning jig's fence. You could position this block higher, if you find it more comfortable.

Crosscut-box guard

A shopmade sliding crosscut box that rides in the tablesaw's miter slots is great for trimming and crosscutting long boards or moldings. And adding a guard is the perfect way to make this sliding jig safer to use. The guard that I made for my crosscut box, as shown in the photo below, is basi-

Crosscuts safe and simple—A three-sided box over the line of the cut reduces the chance of accidental blade contact on the author's crosscut jig. The box, with $\frac{3}{8}$-in. wood sides and a $\frac{1}{8}$-in. Plexiglas top, is held in place at one end by two cleats and rides up and down between them. An exit block guards the blade at the end of the cut.

cally an inverted U-shaped channel that rests on top of the stock over the line of cut, preventing hands from reaching into the blade. This design is very similar to the clear plastic guard that Kelly Mehler built in his article in *FWW* #89, except that mine was made as a retrofit and has wood sides—I don't miss being able to look through the sides of the guard.

I started building the guard by cutting two $2^1/_4$-in.-wide, $^3/_8$-in.-thick wood sides and a $3^1/_2$-in.-wide, $^1/_8$-in.-thick Plexiglas top, all slightly shorter than the front-to-back dimension inside my crosscut box. I then nailed sides and top together with #16 brass escutcheon pins through holes drilled in the plastic. Because the guard was retrofitted to my crosscut box, I couldn't cut grooves for the ends of the guard to slide in, as in Mehler's design. But for a smaller (12-in. capacity) box like mine, two narrow guide strips tacked on the inside of the box's front support are adequate to keep the guard in place and let it ride up and down. Chamfering and rounding the ends and edges of the wood sides makes the guard slide up and down easily. To shield the blade where it exits the crosscut box, I added a rear guard that is a variation on the one for the box-joint jig described previously. In this case, I simply glued and screwed on a wood block to sheathe the blade.

Sliding miter-carriage guard

Many woodworkers like to cut miters on the ends of moldings, picture frames and other trim using a carriage with twin 45° fences, which slides in the tablesaw's miter-gauge slots. When you use this type of jig, you hold the workpiece against the fence during the cut, and your fingers often come close to the blade. And as you finish the cut, the blade exits between the fences, not far from where your thumbs are wrapped over the top of the fences. It's an operation that begs for a guard.

To add protection to my sliding miter jig shown in the photo above, I cut a triangular block from some scrap 2x4 I had around the shop and glued and screwed it to the jig's baseplate just behind the intersection of the fences. This block acts as an exit guard and

Miter shield—A triangular piece of 2x stock serves as an exit block as well as a mounting surface for the Plexiglas blade guard on this sliding miter-carriage jig.

a mounting surface for a clear blade guard. The back end of this blade guard, a 5-in. by 12-in. piece of $^1/_8$-in. Plexiglas, is screwed to the top of the block, and the front end is screwed to a wood strip nailed to the miter jig's front cross support. To complete the safety treatment, I clamp a stop block to the saw table to prevent the blade from cutting through the exit block.

Two resawing guards

Probably one of the most dangerous operations to perform on an unguarded tablesaw is resawing, for two reasons: First, the blade is usually raised to or near its full height. If there's a slipup, you are exposed to more harm than with any other tablesaw operation. Second, there is maximum surface area contact between the wood and the blade. If the wood distorts and binds between the fence and blade (or the kerf closes up and pinches the blade), the workpiece is kicked back with the full force of the saw. These are two excellent reasons to invest a few minutes and a couple of pieces of wood to protect yourself against disaster.

Resawing reconsidered—A chunk of 4x4 screwed to a stick is all that it takes to keep the stock vertical and the blade safely hidden while resawing. If you resaw often, you can screw the guard block directly to a dedicated throat plate.

I've come up with a pair of guarding devices for resawing. Both are simple, but effective. These jigs serve two purposes: They keep the board upright during the cut, and they keep your hands from coming anywhere near the blade.

The first is a clamp-on guard, as shown in the photo above. It consists of a 12-in.-long block of 4x4 lumber with a 2x2 stick screwed to one side. At 3^1/$_2$ in., the 4x4 is thicker than the depth of cut of most 10-in. tablesaws (if your sawblade rises higher, use a thicker block). The block is positioned over the throat plate, just far enough to the left of the blade to allow the stock to feed past. Because the resawn stock will have to be planed anyway, you can set the guard for a fractionally loose fit to account for the distortion caused when the workpiece is cut. The 2x2 stick should be made long enough to center the 4x4 with respect to the blade arbor.

To use the clamp-on resaw guard, set the rip fence, lower the blade into the table and put a piece of stock in place above the blade. Then position the block so it's over the throat plate and snugged up to the workpiece. Secure the end of the stick to the saw table with a C-clamp.

If you do a lot of resawing, you might want to make the second style of guard, which incorporates a dedicated throat plate. On this device, the wood block is attached directly to a replacement throat plate. In addition to providing protection like the clamp-on guard, this version enables you to raise the sawblade through the blank plate for a close fit that supports narrow workpieces right next to the blade. And it prevents the leading edge of the work from hanging up.

Make the replacement throat plate from plywood, particleboard or Masonite that's the same thickness as the original plate. The easiest way I've found to shape the new plate is to use the factory throat plate as a template. I cut out a slightly oversized blank on the bandsaw, attach the factory plate to it with Scotch brand 924 Adhesive Transfer Tape (available in ^{1}/$_2$-in. and ^{3}/$_4$-in. widths from University Products, 517 Main St., Holyoke, Mass. 01041; 800-628-1912) and then trim the new one to size using a piloted, flush-trimming router bit. Once the new plate fits snugly in your saw, screw on the block from below. I keep a couple of these dedicated throat plates handy—one for resawing 4/4 stock and one for 8/4. You can cut slots instead of holes for the screws through the replacement blank to permit adjustment for resawing boards of various thicknesses.

When working with either style of resaw guard, use a push stick to feed the end of the stock through the gap between block and fence—even if the blade is buried in the wood. If resawing must be done in two passes, set the blade height to slightly less than half the width of the board. The board is easily snapped apart after the second pass, and the small unsawn strip down the center of each resawn half can then be planed off. Incidentally, you can also use a similar guard—with a block that's not as high— when ripping narrow strips to width.

Hold-down cove-cutting guard

Passing your hands directly over the blade is dangerous, even if the blade is buried in a thick workpiece—the stock might be kicked back, suddenly exposing the blade. In tablesaw cove cutting, you have to keep constant downward pressure on the workpiece to get good results, so this danger is always present.

My cove-cutting guard, as shown in the photo below, is attached directly to a clamp-on fence, which guides the workpiece across the blade. The guard employs a featherboard-style hold-down over the blade. The hold-down prevents fingers from getting near the blade while keeping the stock flat on the table. And because the hold-down is firmly positioned, it does a better job of flattening the stock than your hands can. The only thing better than a guard like this is a power feeder, which will keep the stock flat on the table and your hands safely away from the blade while feeding the piece for you.

I made the cove-cutting fence from straight-grained stock; I used a $1^3/4$-in.-wide, $1^1/2$-in.-thick piece of Douglas fir. A block of wood $1^3/4$x3x4 is screwed to the top of this fence. Its position along the fence varies depending on the angle of the fence,

which is determined by the desired cove profile (for more on cove cutting, see "Coves Cut on the Tablesaw," *FWW* #102, p. 82). I cut the featherboard from a $4^1/2$-in.-long, 3-in.-wide, 2-in.-thick block and cut the feathers on the bandsaw, making each one about $3/32$ in. thick. Then I attached the featherboard to the fence block with a $3/8$-in.-dia. carriage bolt.

To use the device, clamp the fence to the saw table to the right of the sawblade with the guard centered over the blade. With the sawblade lowered into the table, place the workpiece under the featherboard. Pivot the featherboard until it exerts enough pressure on the piece to press it flat, but not so much that the workpiece is difficult to feed. Depending on the thickness of the work, you may have to relocate the hole for the carriage bolt in the fence block. Finally, clamp a secondary fence to the saw table to keep the work from wandering away from the main fence during cove cutting. As you make each pass over the blade (the blade should only cut about $1/16$ in. deep each pass), use the next workpiece or a piece of scrapwood the same width as the workpiece to push the end of the work under the featherboard.

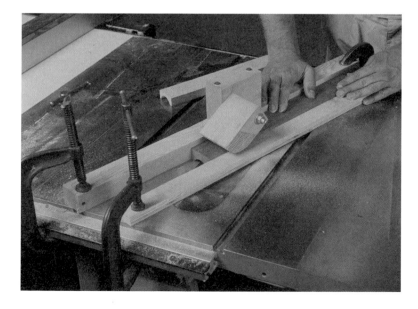

Wide featherboards are excellent for coving—They exert downward pressure over the cutting area while keeping hands from coming near the blade.

ONE-STOP CUTTING STATION

by Ken Picou

Tablesaws are excellent for ripping stock, but the standard miter gauge that comes with most tablesaws makes them mediocre at best for crosscutting material or cutting joinery. But by making a simple sliding-crosscut box and a few accessory jigs, you can greatly increase the accuracy and flexibility of your saw and turn it into a one-stop cutting station, capable of crosscutting, tenoning and slotting.

The system I've developed consists of a basic sliding-crosscut box with a 90° back rail, a removable pivoting fence, a tenoning attachment and a corner slotting jig, for cutting the slots for keyed miter joints (see the photo below). This system is inherently

Making a crosscut box more versatile— An accurate sliding-crosscut box makes a good base for cutting accessories, including this corner-slotting jig. This jig mounts or dismounts in seconds and makes for strong miter joints in picture or mirror frames and in small boxes or drawers.

safer and more accurate than even the most expensive miter gauge for several reasons. First, it uses both miter slots, so there is less side play than with a miter gauge. Second, the work slides on a moving base, so there's no chance of the work slipping or catching from friction with the saw table. Third, the long back fence provides better support than a miter gauge, which is usually only 4 in. or 5 in. across. Fourth, the sliding-crosscut box is big, so angles can be measured and divided much more accurately than with a miter gauge (the farther from its point of origin an angle is measured, the greater the precision). Finally, the sliding crosscut box is a stable base on which to mount various attachments, such as a tenoning jig or a corner slotting jig, which can greatly expand the versatility of the tablesaw.

Building the basic crosscut box

I cut the base of my sliding-crosscut box from a nice, flat sheet of $^1/_2$-in.-thick Baltic-birch plywood, and then I make it a little bit wider and deeper than my saw's tabletop. A cheaper grade of plywood also would be fine for this jig, but I decided to use a premium material because I wanted the jig to be a permanent addition to my shop.

The runners that slide in the tablesaw's miter-gauge slots can be made from any stable material that wears well. I prefer wood to metal because wood works easily, and I can screw right into it. I usually use hard maple, and I've never had a problem. Using a long-wearing, slippery plastic such as an acetal (Delrin, for example) or ultra-high molecular-weight (UHMW) plastic is also a possibility. (For more on using plastics for jigs and fixtures, see *FWW* #105, pp. 58-61.)

I start with a maple board of sufficient length that is at least as wide as three or four runners are thick. I plane this board, taking off minute increments with each pass, until it slides easily on edge in one of the slots but isn't sloppy. Once the fit's right, I rip the runners from this board, setting the fence on my tablesaw to just under the depth of the miter-gauge slot. Then I drill and countersink them at the middle and near both ends (I check the dimensions of the Baltic-birch base to make sure I drill the screw holes so they'll fall near the edges of the base). I usu-

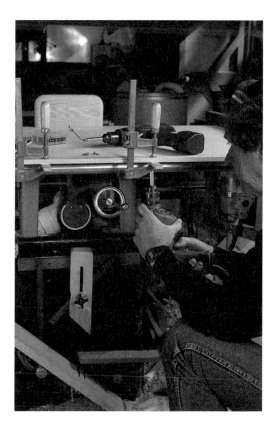

ally drill a couple of holes near each end as insurance in case a screw drifts off when I'm screwing the runners to the base.

Next I crank the sawblade all the way down below the table and lay the runners in the miter gauge slots. I position the base so that its back edge is parallel to the rear of the saw table and the front edge overhangs by a couple of inches. I clamp the runners to the base in the front. I drill pilot holes in the plywood from below using a Vix bit (a self-centering drill bit available through most large tool catalogs) placed in one of the countersunk holes in the runners. Then I screw up through the runners into the base. When I've done both runners at the front of the saw, I slide the base back carefully and repeat at the rear (see the photo above). I check for binding or wobble by sliding the base back and forth a few times. If the fit is less than ideal, I still have four more chances (the extra screw holes I drilled at both ends of each runner) to get it right. If the fit is good, I drill pilot holes with the Vix bit and screw the runner to the base in the middle, taking care not to let the runner move side to side. I also trim the runners flush with the front and back of the crosscut box.

Sliding crosscut box and accessory jigs

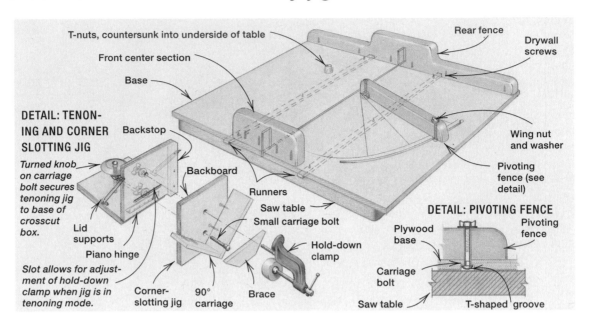

T-nuts, countersunk into underside of table

Front center section

Base

DETAIL: TENON-ING AND CORNER SLOTTING JIG

Turned knob on carriage bolt secures tenoning jig to base of crosscut box.

Backstop

Lid supports

Piano hinge

Slot allows for adjustment of hold-down clamp when jig is in tenoning mode.

Backboard

Runners

Saw table

Small carriage bolt

Corner-slotting jig

90° carriage

Brace

Hold-down clamp

Rear fence

Drywall screws

Wing nut and washer

Pivoting fence (see detail)

DETAIL: PIVOTING FENCE

Plywood base

Pivoting fence

Carriage bolt

Saw table

T-shaped groove

If the fit's a bit too snug at first, use will tend to burnish the runners so that they will glide more easily. If, after some use they're still a little snug, you can sand the runners just a bit and give them a coat of paste wax. That will usually get them gliding nicely.

Building accuracy into the jig

An inaccurate jig is useless, so it's essential that assembly of this jig be dead-on. Fortunately, this isn't difficult; it just takes a little time and patience.

I made both the back fence and the front center section from straight-grained red oak, but any straight-grained hardwood will do (see the drawing above). I make sure the center portions of both pieces are built up high enough to provide $1^1/_2$-in. clearance with the blade cranked up all the way.

The front section helps keep the table flat and prevents it from being sawn in half. Because this front section is not a reference surface, its position isn't critical, so I screw it on first.

Then I mount the rear fence about $^1/_4$ in. in from and parallel to the back of the Baltic-birch base. I clamp the fence to the base and drive one screw through the base, which I've already drilled and countersunk, into the fence a couple of inches to the right

of where the blade will run. This provides a pivot point, making it easier to align the rear fence to the blade.

I remove the clamp, raise the blade up through the base and cut through the front section and the base, staying just shy of the rear fence. So far, there's only one screw holding the rear fence in place. To set the rear fence permanently and accurately at 90° to the blade, I place the long leg of a framing square against the freshly made kerf (saw is *off*) and the short leg against the fence. With the fence flush against the square, I clamp the fence on an overhanging edge and do a test-cut on a wide piece of scrap. I check this for square with a combination square and adjust the position of the fence as necessary. When I've got it right, I put another clamp on the fence near the blade on the side opposite my one screw. Then I drill, countersink and screw through the base into the fence right next to the clamp, and I check the fence's position again to make sure screwing it to the base didn't pull it off the mark (see the photo on the facing page). I also make another test-cut, and as long as it's still good, I screw the fence down near the ends and the middles on both sides of the blade (see the drawing above). If the second cut is not a perfect 90°,

Checking and rechecking for a perfect 90°, both with a square and with test-cuts, is time well-spent. The accuracy of the whole cross-cut box and all jigs that mount to it depends on getting the relationship of rear fence to blade just right.

then I'll fiddle with the fence until the cut is perfect before screwing it into position permanently. Time spent getting the fence right is time well spent. If, for aesthetic reasons, you want the rear of the base to be flush with the fence, you can trim the base flush with a bearing-guided, flush-trimming router bit. Either way, the performance of the crosscut box will be unaffected.

Anything from a small wooden handscrew to a fancy commercially made stop will work as a stop block for this fence. A self-stick ruler can be added to the fence or table.

A pivoting fence

I wanted a pivoting fence for making angled cuts, but I also wanted to be able to remove the fence quickly when I need to cut wide boards. I accomplished this first by setting a T-nut for the pivot point into the underside of the jig's base about 6 in. forward of the fixed fence. Then I routed an arc-shaped track for a carriage bolt at the end of the fence (see the drawing on the facing page). The arc runs from 0° to a bit more than 45°, and there's a plunge-routed hole just below the 0° point through which the carriage-bolt assembly can be lifted out to remove the fence. I marked two common angles ($22^{1}/_{2}$° and 45°) onto the jig for

quick reference using a large protractor and transferring that angle to a bevel square and then to the plywood. These angles can also be checked and fine-tuned by cutting them, setting the resulting blocks together and checking for 90° with an accurate square.

A slotted screw and washer secure the fence at its pivot point but allow the fence to move, and a wing nut (with washer) fixes the angle of the fence at its outboard end. As with the fixed fence, a stop block may be as simple or sophisticated as you like.

An adjustable tenoning jig

A simple hinged jig that uses the rear fence as a reference surface will allow you to cut both regular and angled tenons, rabbets and angled edges accurately and without too much fuss. I built this jig also from Baltic-birch plywood. I crosscut it in the basic jig and routed the slots in it on my router table.

To attach the hinges accurately, I indexed both halves against the fixed rear fence, set a length of piano hinge in place and used a small carpenter's square to align the hinges (see the top left photo on p. 128). Then I drilled screw holes using the Vix bit and screwed the hinge on.

Rear fence helps align jig's hinge—Using the rear fence as his reference, the author aligns the tenoning jig's hinge with a square. The Vix bit ensures that the screw holes are centered, so the screws will go in true and the hinge will be straight.

Quick, accurate tenons, even in large boards, are easy with the author's hinged tenoning jig. A hold-down clamp grabs the workpiece securely and accommodates almost any size workpiece. The big footprint of the tenoning jig's base anchors it securely to the base of the crosscut jig below. The jig is also useful for cutting long miters and angled tenons.

Setting angles accurately can be done quickly with a miter square or a bevel square. By setting the angle both fore and aft in the tenoning jig, you can be sure the angle will be true across the face of the jig.

A small shopmade (turned) knob at the end of a carriage bolt secures the tenoning jig to a T-nut in the underside of the crosscut box's base. The fixed rear fence ensures that the face of the tenoning jig stays parallel to the blade. Two brass lid supports hold a set angle securely (see the bottom left photo above). And a hold-down clamp travels in a slot in the upper portion of the jig, allowing me to hold almost any size workpiece securely (see the photo above right).

Corner-slotting jig

Attaching directly to the tenoning jig, the corner-slotting jig is easy to build and simple to use. I screwed two scrap boards to a backboard to form a 90° carriage positioned at 45° to the base of the crosscut box (see the drawing on p. 126). I cut a brace to fit up a few inches from the corner of the 90° carriage and across whatever it is I'm slotting. A hole through the backboard permits a hold-down clamp to bear upon the brace, distributing the pressure of the clamp.

In use, I slide the workpiece into place, then the brace and then I tighten the clamp. The jig feels solid and works well.

A TABLESAW SLED FOR PRECISION CROSSCUTTING

by Lon Schleining

Miters

Crosscuts

Tenons

Crosscutting with a standard tablesaw miter gauge can be frustrating, inaccurate, even hazardous. Adding an extended fence helps, but the miter gauge still will be limited and imprecise. Don't bother with it. Instead, take the time to make a super-accurate, super-versatile and far safer crosscut sled.

A crosscut sled is a sliding table with runners that guide it over the saw in the miter-gauge slots. It has a rear fence set perpendicular to the line of cut to hold the workpiece. Because it uses both miter slots, the sled is remarkably and reliably accurate. It also easily accepts any number of stop blocks, auxiliary fences and templates, allowing miters, tenons and many other specialty cuts. Nearly every small commercial shop I know uses some variation of this sled. I use mine primarily to square the ends of 12-in.-wide stair treads.

Your sled should fit your work. There's no sense in making a huge, unwieldy sled if you'll use it mostly to cut 3-in. tenons. The one I use is 30 in. wide and 21 in. deep. It's capable of crosscutting a board up to 2 in. thick and 18 in. wide (see the photo at left on p. 129). With a miter template (see the sidebar below), the sled can cut a 45° miter on the end of a 3-in.-wide board. The rear fence is 5 in. high in the middle, 2¹/₂ in. high on the ends. Though I rarely crosscut a board thicker than 2 in., the fence needs to be at least 4 in. high to accommodate the height of the sawblade. The extra fence height also supports workpieces on end when I cut tenons.

Start with a solid platform of Baltic-birch plywood

I build jigs like this from what I call not-yet-used materials (some call it scrap). I used void-free ¹/₂-in. Baltic-birch plywood for the platform. Baltic birch is often mistaken for Finnish birch—its waterproof and much more costly cousin. Baltic birch is not as high quality, but for the price (about a dollar per square foot), it's perfect for making stable, durable jigs. But any plywood you have around the shop will probably work fine as long as it's flat.

The first step is to cut the platform to size. Make the platform as square as you can get it. You can check for square by measuring diagonally across the corners: The measurements should be the same across both corners. But before you make the sled, it's a good idea to make sure your tablesaw is tuned up.

From 90° to 45° cuts with a simple template

With this template, you'll be able to make accurate miter cuts on your tablesaw. The template is nothing more than a piece of Baltic-birch plywood with two sides at 90° to each other and a back side that registers against the rear fence of the sled. This template sits far enough forward so that long workpieces clear the ends of the rear fence.

There are any number of ways to make such a shape. I used the opportunity to test the accuracy of my sled. First I laid out and rough cut the template from a corner of a sheet of plywood and got one of the sides straight on a jointer. This can also be done on the sled by aligning the edge over the sawkerf and nailing the tem-

plate to the sled (don't let the nails go all the way through). I then cut the opposite side at 90° to the first using the rear fence.

To cut the base at 45° to the two sides, I cut to the layout line on the base by aligning it over the kerf and nailing the template to the sled. I've rarely gotten a base perfect the first time.

To find out which way it's out, I center the point of the template on the sawkerf and align the base against the rear fence. Then I scribe its outline on the sled. I flip it over and check it against the scribe marks. If it sits perfectly between the lines, I'm on the money. If not, I recut the back of the template as required. Finally, I attach it to the sled with a few screws, make some trial miters and adjust accordingly. —L.S.

A basic crosscut sled

Tailor the size of the sled to fit the work you do. The crucial features are a rear fence perpendicular to the line of cut and runners that slide easily without stop.

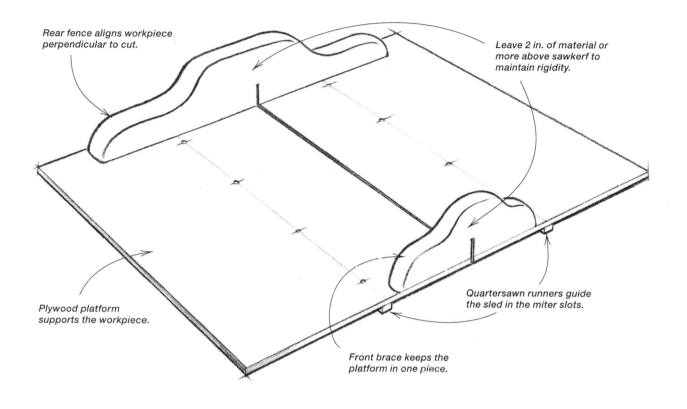

Rear fence aligns workpiece perpendicular to cut.

Leave 2 in. of material or more above sawkerf to maintain rigidity.

Plywood platform supports the workpiece.

Quartersawn runners guide the sled in the miter slots.

Front brace keeps the platform in one piece.

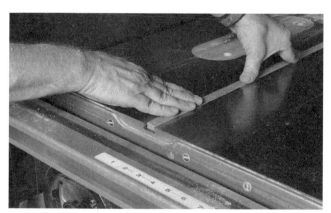

Start with the runners—Runner stock should slide freely in the miter slots (right). Finished runners should be just below the level of the table (above).

For the sled to perform well, your saw's blade must be precisely parallel with the miter-gauge slots, and the table must be flat (for more on tuning up your tablesaw, see *FWW* #114, pp. 60-64).

Quartersawn hardwood runners for smooth sliding

I prefer to make runners from oak, instead of buying steel ones, because I can control their fit in the miter slots. Wood runners pose a few problems, however, that should be taken into consideration. Expansion from seasonal humidity can cause them to bind in the miter slots, so I choose the material and its grain orientation carefully. They also need to be milled precisely.

Start with a close-grained flatsawn maple or oak board. Mill the thickness of the board to the width of the slot using a planer. Test the fit as you go, planing off a little material at a time. It should slide easily in the slot, but without slop (see the photos on p. 131).

Next rip two runners from the board to a thickness slightly less than the depth of the miter slots, then cut them to length. By ripping strips off a flat-grained board, you have made quartersawn runners, which will be very stable. The idea is to make runners that don't rub against the bottom of the slots and raise the sled off the table, but that still engage as much of the miter slot as possible.

The first construction step is to fasten the runners to the platform. To make sure they are right where they should be, attach them while they're in the miter slots. Lower the blade out of the way, and center the platform on the table, using the rip fence to keep the platform square on the runners (see the photo at left below). Lay out the holes for the screws so they're centered on the runners, and drill them in the platform only. The screws should pass freely through the holes in the plywood.

The size of the drill bit you choose for the pilot holes in the runners is very important.

Attach the platform to the runners—Use miter slots to align runners under the platform. The rip fence keeps the platform square and centered while you lay out (above) and drill the pilot holes (right). To avoid splitting the runners, the holes should be slightly larger than the shank diameter of the screw.

Thin runners will bulge or split if the pilot hole is too small. Even a small bulge will make the runner bind in the miter slot. The holes should be slightly larger than the shank diameter of the screw. I use a dial caliper to measure the shank, and then I select the correct drill bit. On this sled, I used ⁵/₈-in.-long #8 screws that have a shank diameter of 0.122 in., so a ¹/₈-in. drill bit (0.125 in.) was perfect.

First drill just one pilot hole in each runner, and insert a screw in each. These screws keep the runners firmly in place while you drill the other pilot holes. Remove the two screws, deburr all the holes, apply a small bead of glue to the runners and screw the platform to the runners. Clean off any glue that might have squeezed out.

Now take your incomplete sled for a test drive: move it back and forth in the miter slots to see if it runs smoothly. It's easy to tell just where the oak runners are binding because they'll be shiny and gray from rub-bing against the sides of the steel slots. While the glue is still soft, it is possible to move the runners slightly. You should only be concerned at this point with how smoothly the platform slides.

Make front brace and rear fence

The front brace's only job is to keep the platform in one piece. It doesn't much matter what size or shape it is (I add some gentle curves to mine) as long as it is a few inches higher than the sawblade's maximum cut—about 2 in. above the platform. I made this brace from 1¹/₄-in.-thick red oak, 3³/₄ in. high, and about as long as the width between the miter slots. Shape it, smooth it, glue and screw it to the front of the table from the underside of the platform.

This is also the time to make the rear fence. I used some 2-in.-thick white oak 5 in. wide and 23 in. long. The rear fence should be pretty stout to hold the sled table together. If you don't have 8/4 lumber, lami-

Use the kerf to square the fence— Don't cut the sled in half. After you attach the front brace (above), cut only two-thirds of the way through the platform (right). The kerf is a reference to set the rear fence.

Square the fence to the sawkerf. Check the fence's alignment from both sides of the kerf. Attach the fence with only two screws before you make trial cuts.

just made. Use an accurate framing square to align it, checking from both sides of the fence. Now drill two center pilot holes (of four total) into the fence, and install the screws from the bottom side.

Before you can attach the rear fence once and for all, make some trial crosscuts and check the results. The position of the fence will almost certainly need fine-tuning. It's easy to rotate the rear fence back and forth a little with hammer taps or a bar clamp, even with the two screws snug. This is where patience is important. Keep making test cuts and adjusting as necessary until the cut is perfectly square. Don't, however, cut all the way through the platform at this time. Leave just enough plywood at the rear of the platform to hold the sled together; if you cut all the way through, the rear fence will be harder to align.

Attach the rear fence and make more trial cuts

When the sled makes true 90° crosscuts, it's time to attach the rear fence permanently. Clamp a long 4-in. by 4-in. block to the sled platform so that it fits tight against the rear fence. It will keep the fence's place. Remove the two screws that are temporarily holding the fence. Apply glue and reinstall the fence with the rest of the screws. Carefully check its position against the block. Remove the clamps and the block, and immediately make a trial cut, still without cutting all the way through the platform.

Adjust the fence if necessary with hammer taps or clamps. Even though the sled is screwed and glued together at this point, it's still possible to make fine adjustments, but only for a few minutes after glue-up.

Before you spend too much time admiring your handy work, sand all the sharp edges and coat the bottom with a lubricant such as spray silicone or TopCote. Even then, you're not done. You still have guide blocks and templates to make. They will let your sled cut perfect tenons and miters.

nate two 4/4 pieces together. Make sure the board is perfectly straight on the inside face, and square with the edge that will be attached to the platform.

Keeping things square becomes critical when you attach the rear fence. The most important thing to remember when making a sled is that, for the cut to be square, the rear fence must be square to the line of cut. If it's not, you have a useless sled.

Before you attach the rear fence, put the sled on the saw, raise the blade slightly above the thickness of the platform and cut through the platform about two-thirds of the way from back to front, being very careful not to cut all the way through the platform (see the bottom photo on p. 133). Drill and countersink the holes in the platform, then securely clamp the fence to the platform so that it is square to the cut you

SLIDING TABLE FOR ANGLED TENONS

by William Krase

Angled tenons can be difficult to cut, but Krase's system greatly simplifies the process. The workpiece seats securely against the wedges at the junction of the crossfeed and sliding table, while the sliding table guides the whole affair through the blade.

Angled tenons—some compound—were used almost exclusively in the construction of this walnut chair, stool and side table. Legs on two of these pieces splay in two directions, requiring slightly angled tenons at both ends of apron pieces, stretchers and seat supports.

Lots of furniture—especially pieces intended to accommodate the human body—require joints that are not square. Chairs may have as many as 16 such joints, some of which are compound (angled in two planes). That's why chairs can be difficult. They don't have to be.

With my addition of a crossfeed box to Kelly Mehler's sliding table (*FWW* #89, p. 72) and the use of purpose-made wedges, you can cut even compound-angled tenons quickly, accurately, time after time (see the photo on p. 135). The wedges establish the tenon angle while the crossfeed box positions the workpiece to get the correct length, width and thickness of tenon.

I arrived at this method of cutting angled tenons because I wanted to make the stool in the photo above. Since then, I've used it on four more pieces of furniture—over 60 angled joints in all. Though now I wish I'd made the sliding table and crossfeed box of a better material, I've been completely satisfied with both the apparatus and the results.

I used regular particleboard (the kind often used for floor underlayment) for the sliding table's base and for the crossfeed box (see the drawing on the facing page for critical dimensions and construction information). Particleboard is what I had handy, but if I

were to build another, I'd use medium-density fiberboard (MDF) or a good-quality birch plywood instead. Particleboard seems to be susceptible to changes in humidity, resulting in some binding whenever the humidity becomes extreme.

I make wedges for projects as I need them. They must be long enough to support the workpiece securely in the upright position. I've found that 1-ft. sections of 2x stock work well.

To make the thumbscrews that fasten the crossfeed box to the sliding table, I bought a length of $^{1}/_{16}$-in. by $^{1}/_{2}$-in. brass strip (from a hobby shop), cut pieces to size and soldered them into the head slots of slotted brass machine screws. The resulting home-made thumbscrews are oversized, so it's easy to tighten the crossfeed box in place. I use large washers beneath the thumbscrews to prevent them from digging into the crossfeed box.

Cutting tenons

Generally, the first thing I do when cutting angled tenons is to cut the end of the workpiece parallel to what will be the shoulder of the tenon, using the sliding table and wedges. Then, when I position the wedge (or wedges), I make sure the end of the workpiece flushes up against the crossfeed

Sliding table system for cutting angled tenons

The addition of a crossfeed box to a sliding table along with the use of purpose-made wedges make it possible to cut accurate, repeatable angled tenons on a tablesaw in very little time and with a minimum of effort.

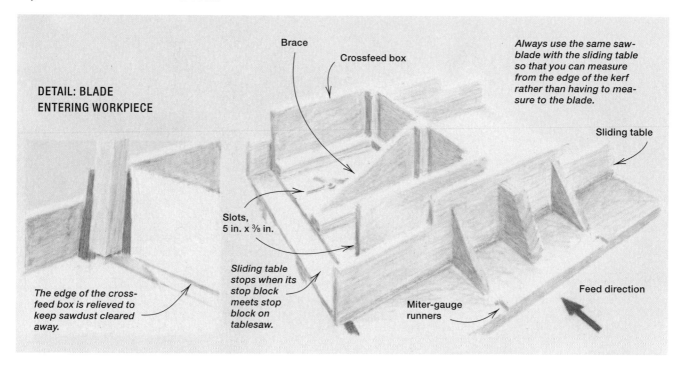

Brace

Crossfeed box

Always use the same saw-blade with the sliding table so that you can measure from the edge of the kerf rather than having to measure to the blade.

DETAIL: BLADE ENTERING WORKPIECE

Sliding table

Slots, 5 in. x ⅜ in.

Sliding table stops when its stop block meets stop block on tablesaw.

The edge of the cross-feed box is relieved to keep sawdust cleared away.

Miter-gauge runners

Feed direction

---

box (for cutting shoulders) or the base of the sliding table (for the cheeks). This helps orient the workpiece and minimizes the chance of my ending up with an expensive piece of kindling. That's happened only once using this jig, when I measured to the wrong side of the sawblade.

Tenons angled in one plane require one wedge; compound-angled tenons require two. I use the same wedges for cutting both the shoulders and cheeks. The wedges just have to be manipulated to reposition the workpiece properly with respect to the blade—in practice the orientation is obvious. As a rule, I cut the shoulders first

and then the cheeks. This creates a crisp shoulder, makes cutting the cheeks easier and minimizes the chance of pinching the blade with the small offcuts.

With the workpiece bearing against two surfaces oriented 90° to each other and with the force of the blade only serving to seat the workpiece more securely, I'm comfortable handholding the workpiece. If it makes you feel safer or more secure, by all means, use a clamp, but just be sure the clamp doesn't vibrate loose and fall into the blade.

TAPER JIG IS SIMPLE, SAFE AND EFFECTIVE

by Gary Rogowski

I make my living as a woodworker, so I need to spend more time making furniture than making jigs or fixtures. My approach to jig making is no-nonsense: What's going to give me accurate, consistent results, safely and quickly? Which brings me to my taper jig.

It's a dedicated jig (that is, it's for one taper only; it's not adjustable), so it isn't as versatile as it might be. But it more than

Dedicated taper jig ensures safe, consistent results. With stops both front and back, this jig captures the leg snugly, keeping it from vibrating or moving as it's cut.

After the first taper cut, turn the leg so the cut faces up. This keeps a jointed face on the saw table and a square end against each stop for the second cut.

Basic taper jig

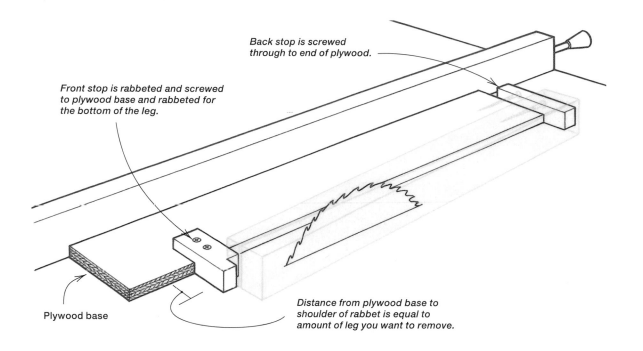

Back stop is screwed through to end of plywood.

Front stop is rabbeted and screwed to plywood base and rabbeted for the bottom of the leg.

Plywood base

Distance from plywood base to shoulder of rabbet is equal to amount of leg you want to remove.

makes up for that in safety. The leg is captured front and back rather than just in the back as is the case with most adjustable jigs. I've made three more harvest tables since the first one, and I've been able to depend on this jig for consistent tapers.

To make the jig, I ripped a piece of plywood about 6 in. wide (the width isn't important, but this feels about right to me) and about 3 in. longer than the leg I'm tapering. At the back edge of this plywood, I screwed a fairly wide block of wood, so I'd have a sturdy back stop. I made sure this block was flush with the bottom edge of the plywood and was sticking out far enough from the edge of the plywood to act as a stop for the leg (see the drawing above). The top of the leg will fit against this stop.

The leg bottom is the first part to enter the sawblade. It needs to be set out away from the plywood base of the jig a distance equal to the amount you want to remove. For this table, I wanted to taper the legs from $2^{1}/_{2}$ in. at the top down to 1 in. at the bottom, so I needed to push the leg bottom out $1^{1}/_{2}$ in. from the edge of the plywood.

To do this, I made another stop that positively located the leg $1^{1}/_{2}$ in. from the edge of the plywood. I cut two rabbets in this stop at 90° to each other, one indexing against the plywood and the other securing the bottom of the leg. I screwed this stop onto the plywood.

If the leg doesn't fit snugly between these stops, glue on a piece of sandpaper or some other shim to make sure it does. When you run those legs through the blade, with over $2^{1}/_{2}$ in. of blade protruding from the table, you want to know the leg is positively captured in the jig, not vibrating around.

FIVE

Jigs for Other Tools

Jigs are not the food of the router and tablesaw alone. All other workshop tools can benefit from the guidance potential of jigs. Tools such as the drill press, lathe, planer, and chop saw need jigs and accessories like the tablesaw. They're just a little more specialized and not as adaptable. They consequently attract less attention from jig builders.

In some ways, jigs for these types of tools take a little more brain power to figure out. This isn't to say that you need to be a rocket scientist to invent a jig for one of these tools (although don't be surprised if you find one among the authors). It's simply that the router and tablesaw are relatively easy to adapt to specialized roles with a jig. On the other hand, a planer is already specialized. How would you go about, say, eliminating snipe without pulling the whole thing apart and redesigning internal components? One of the authors found a very simple way.

Most of the other tools, such as the drill press, aren't so much of a challenge, but they still take a fair amount of noodling. Turning a tool into an end boring jig for long pieces isn't as intuitive a task as making a taper jig for a tablesaw, but someone figured it out. And how might you go about making a sliding table for a drill press that turns it into a mortiser? You'll find the answer here as well. A small vacuum clamping table, two jigs to hold bowls on the lathe, and a system for panel storage round out this final collection of articles.

SLIDING TABLE FOR A DRILL PRESS SIMPLIFIES MORTISING

by Mac Campbell

Ideal for mortising, this shopmade sliding table advances a mahogany stile under a fluted stagger-tooth bit chucked in an overarm router. Author Mac Campbell secures the workpiece to the fence with a block of scrap and a C-clamp.

When I need to cut lots of mortises, I like to use a tool that does a consistent job without requiring an involved setup each time. A few years ago, I had to mortise a sizable run of custom chair components, so I picked up a small, used overarm router. To make mortising more efficient, I fitted the machine with a sliding table (see the photo at left), which was quick to build using plywood and standard hardware I had on hand. Adjustable stops make setting the table's horizontal movement straightforward, and the pedal-fed router makes vertical (mortise depth) setup and plunging a cinch.

Rockwell no longer manufactures the machine I have, but similar tools that have a rigid arm supporting a router at 90° to a height-adjustable table are still made. Even without an overhead router, the sliding table can be mounted on a drill press or used for horizontal routing. In addition, the table's indexing (stop) system is especially well-suited to hollow-chisel mortising.

Assembling the table

The sliding table consists of upper and lower plywood pieces connected by pairs of drawer slides and hardwood runners, as shown in the drawing. The table size depends on the tool you are mounting it to and on the length of the drawer slides you're using. The slide mechanisms let the tabletop travel laterally over the base, which is bolted to the tool. The larger the table (mine is 20 in. long), the more stable the setup will

142

Sliding table for mortising

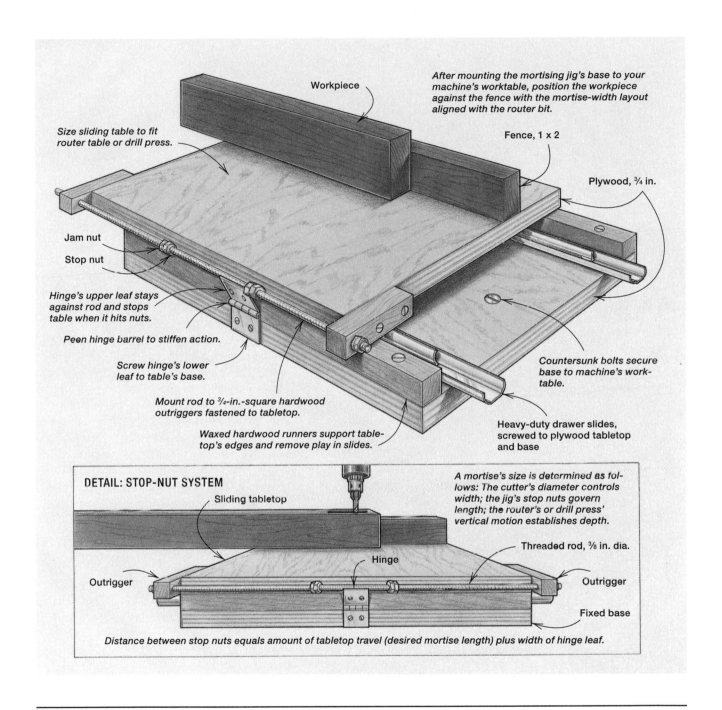

Workpiece

After mounting the mortising jig's base to your machine's worktable, position the workpiece against the fence with the mortise-width layout aligned with the router bit.

Size sliding table to fit router table or drill press.

Fence, 1 x 2

Plywood, ¾ in.

Jam nut

Stop nut

Hinge's upper leaf stays against rod and stops table when it hits nuts.

Peen hinge barrel to stiffen action.

Screw hinge's lower leaf to table's base.

Mount rod to ¾-in.-square hardwood outriggers fastened to tabletop.

Waxed hardwood runners support tabletop's edges and remove play in slides.

Countersunk bolts secure base to machine's worktable.

Heavy-duty drawer slides, screwed to plywood tabletop and base

DETAIL: STOP-NUT SYSTEM

Sliding tabletop

A mortise's size is determined as follows: The cutter's diameter controls width; the jig's stop nuts govern length; the router's or drill press' vertical motion establishes depth.

Threaded rod, ⅜ in. dia.

Hinge

Outrigger

Outrigger

Fixed base

Distance between stop nuts equals amount of tabletop travel (desired mortise length) plus width of hinge leaf.

be. A piece of 1x2 secured across the width of the tabletop serves as a fence.

The drawer slides are the heavy-duty variety intended for file cabinets. They are equipped with a metal track that's designed to be mounted to a cabinet side and a similar track for the drawer side. A ball-bearing carriage runs between the two tracks. My slides are 20 in. or so long and allow about 25 in. of travel. These slides have virtually no play and don't depend on gravity to keep everything aligned. A little judicious tinkering will remove the stop on each slide that prevents the drawer from over-closing, allowing the mortising jig to extend in both directions.

The hardwood runners support the tabletop's edges and eliminate any play that might develop at the extremes of travel when there is a relatively short length of the bearing carriage between tracks. The runners should be thick enough to fill the gap between the sliding table and the base. Waxing the runners lets the jig move smoothly.

Installing adjustable stops

A length of $^3/_8$-in.-dia. threaded rod installed along the front of the upper table (see the drawing detail on p. 143) is the basis for the table's stop system. I bolted the rod to hardwood outriggers on the ends of the table. Nuts threaded onto the rod serve as stops that regulate table movement. Several pairs of nuts act as multiple stops; the second nut in each pair (a jam nut) locks the first in place. A hinge screwed to the plywood base bears against the rod and stops the table when its upper loose leaf hits the nuts. The secret for making the hinge stiff enough for the upper leaf to remain upright against the rod is to hammer on the hinge barrel a little.

Mortising on a sliding table

Once everything is assembled and mounted on the overarm router table, I install a straight-cutting bit and then set it to take a fine shaving off the fence itself. This operation trues the fence to the bit regardless of irregularities anywhere along the line. This, in turn, guarantees that the mortises will be parallel to the face of the stock that's clamped to the fence. When mortising with my overarm router, I use a two-flute, stagger-toothed carbide bit, which is available from Furnima Industrial Carbide, PO Box 308, Barry's Bay, Ontario, Canada K0J1B0; (800) 267-0744. I'm told that these fluted bits will only work at speeds of 17,000 rpm and up. Therefore, using them in a drill press is *not* an option.

To mortise a stile, leg or what have you, first lay out the mortise on the stock. It need not be centered; in fact, many applications work better with an off-centered mortise and tenon. A chair rail, for instance, should have the tenon near its outside face, allowing greater penetration into the leg without cutting into the tenon on the second rail, which enters the leg at 90°. After marking the length of the mortise (tenon width) on the appropriate pieces, I locate the center of the layout lines in approximately the correct position on the jig and then clamp the work to the fence. To ensure uniform depth, I preset the machine's depth stops. Because the bit determines the width of the mortise (tenon thickness), cutting mortises is just a matter of aligning the stops with the marks on the stock and routing the slot.

I raise the machine's table to take about $^1/_8$ in. each pass while I slide the jig (with workpiece clamped to the fence) back and forth under the cutter (see the photo on p. 142). Each mortise takes only 10 or 15 seconds, and changing the stock takes about the same. A series of mortises for 20 frame-and-panel doors takes only a half hour or so, plus, perhaps, five minutes for setup.

END-BORING JIG FOR A DRILL PRESS

by Jeff Greef

I've built a fair number of custom doors and windows in which I've joined the stiles to rails with dowels. Until recently, I relied on doweling fixtures to position holes. Although fixtures are quick to use, I found them lacking accuracy, particularly for large dowels. The problem is not in locating the fixture's bit guide precisely, but rather it is guaranteeing that the bit drills straight and true.

A horizontal boring machine could solve the problem. But while one of these machines is neither hard to use nor difficult to build, it would eat up precious space in my already cramped shop. Besides, it seemed redundant to buy a motor, bearings and a thrust mechanism when I already had all those things standing in the corner in my drill press.

Boring holes with a drill press is very accurate, but how do you bore holes in the ends of long workpieces? If you turn the press table vertically, clamp a fence to it and secure the work, the setup is still quite limited in terms of making fine adjustments. So with adjusting (and readjusting) in mind, I made an end-boring jig that mounts to my drill press, as shown in the photo at right.

Boring the ends of long workpieces used to present problems for the author until he devised this end-boring jig for his drill press. A pair of platens mounted to the press's table allows adjustments in and out and left and right.

Drill-press mounting logistics

Before you build the jig, you need to figure out how you will mount it. Although each type of drill press may require a slightly different setup, you should be able to adapt the principles I used to mount a jig to your press. First, my jig is designed for floor-model presses. If you have a bench press, you'll have to bolt its base to a workbench with the spindle overhanging the edge. You can extend the jig to the floor as I did (see the drawing below). Second, the jig is made for presses with at least 14 in. of swing to get the depth to the column needed for mounting. Third, the jig is built for presses with tables that both tilt and swivel. By swiveling the table's arm 90° and tilting the table vertically, you can bolt or clamp the jig to it. If your table doesn't tilt and swivel, remove it. Then make a wooden outrigger

End-boring jig assembly

Jig consists of a fixed platen that bolts to the drill-press table and a movable platen, which is made up of three layers.

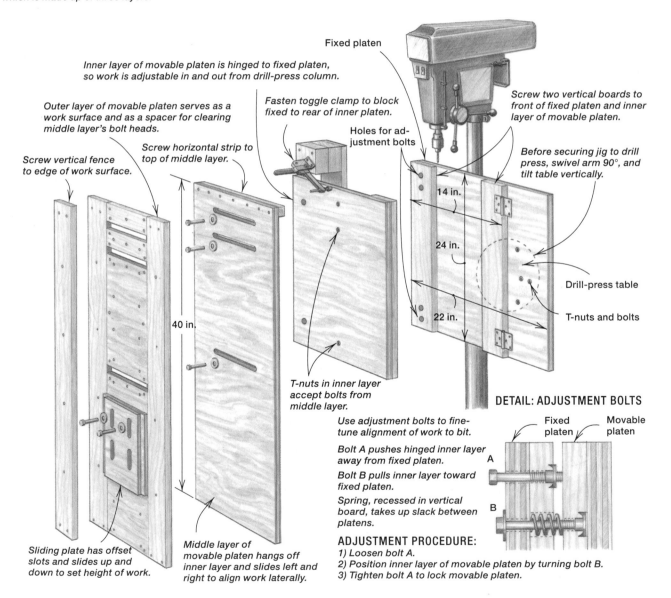

Inner layer of movable platen is hinged to fixed platen, so work is adjustable in and out from drill-press column.

Outer layer of movable platen serves as a work surface and as a spacer for clearing middle layer's bolt heads.

Screw vertical fence to edge of work surface.

Screw horizontal strip to top of middle layer.

Fasten toggle clamp to block fixed to rear of inner platen.

Fixed platen

Holes for adjustment bolts

Screw two vertical boards to front of fixed platen and inner layer of movable platen.

Before securing jig to drill press, swivel arm 90°, and tilt table vertically.

14 in.

24 in.

22 in.

Drill-press table

T-nuts and bolts

40 in.

T-nuts in inner layer accept bolts from middle layer.

Sliding plate has offset slots and slides up and down to set height of work.

Middle layer of movable platen hangs off inner layer and slides left and right to align work laterally.

Use adjustment bolts to fine-tune alignment of work to bit.

Bolt A pushes hinged inner layer away from fixed platen.

Bolt B pulls inner layer toward fixed platen.

Spring, recessed in vertical board, takes up slack between platens.

DETAIL: ADJUSTMENT BOLTS

Fixed platen

Movable platen

A

B

ADJUSTMENT PROCEDURE:
1) Loosen bolt A.
2) Position inner layer of movable platen by turning bolt B.
3) Tighten bolt A to lock movable platen.

with a yoke to clasp the press's column. Mount the jig to the outrigger in line with the spindle.

Designing and building the jig

The boring jig has two main parts: a fixed platen and a movable platen. The fixed platen bolts onto the drill-press table. The movable platen attaches to the fixed one with hinges on one side and adjustment bolts on the other. The hinges allow the movable platen to be positioned in and out from the press's column. The adjustment bolts fine-tune the alignment to the bit. The bolts work in a push-me-pull-you fashion (see the drawing detail on the facing page): One bolt pushes the movable platen away, the other bolt pulls it toward the fixed platen and a spring takes up slack. The movable platen also slides to move work left or right, and an adjustable stop plate sets the height of the work. A toggle clamp secures the workpiece alongside a fixed, vertical fence.

Fixed platen

Because of the jig's $4^1/_2$-in. depth, I turned the press's table a full 90° and pivoted it to vertical before I bolted on the fixed platen, which is just a piece of $3/_4$-in. plywood. I made the platen 22 in. wide to span the distance to the spindle. On the front of the platen, I fastened two vertical boards: one 7 in. to the left of the press' spindle, one 7 in. to the right. I mounted a pair of hinges to the right board, and I recessed T-nuts in the left one to receive the adjustment bolts. Two boards on the back of the movable platen mate with the hinges and bolts.

Movable platen

The movable platen consists of three layers. The outer two layers of the movable platen are long enough to double as a support because the jig and workpiece are suspended from the drill-press table. The inner layer is hinged and bolted to the fixed platen. The middle layer has a strip of wood screwed along the top of it, so it hangs off the inner layer. I made the middle layer so that it can slide left and right; that way, it's easy to precisely bore side-by-side holes in the same plane of a piece. The outer layer is the work surface and also acts as a spacer to keep work clear of the middle layer's bolt heads. I used fender washers under the bolt heads to avoid splintering the plywood.

Vertical fence, work clamp and adjustable stop plate

A vertical fence on the left side of the work surface and a toggle clamp fastened to the rear of the inner layer secure the work. The middle layer can still slide without affecting the clamp. During end-boring, the work rests on a stop plate. The plate is slotted to handle pieces up to 40 in. long, and the slots are offset, so the plate can be flipped for other heights. To set the height of a piece, I just slide the stop plate up or down on two bolts.

Using the jig

Once I've set up the jig and positioned the work, I wedge in a pair of shims between the bottom of the movable platen and the top of the press's base to stabilize the jig (see the photo on p. 145). Even with the platen wedged in place, I can make up to $1/_{32}$-in. corrections using the adjustment bolts.

To break in the jig, I bored holes in the ends of rails on 12 interior doors. The end-boring jig proved a real improvement over my conventional doweling fixtures. Once the work was roughly aligned, it was easy to make fine adjustments on test scraps before boring the actual run of holes. I still keep the old doweling fixtures on hand—but only for those rare situations that the end-boring jig won't handle.

VACUUM POWERED HOLD-DOWN

by Evan Kern

As an instrumentmaker, one of my challenges is planing resawn wood to less than $1/4$ in. thick. My attempts to thickness wood with a conventional planer usually result in hopelessly warped or shattered pieces of wood. Although an abrasive planer can do the job, one of those is well beyond my financial means. And since my needs for thin stock are modest, I bought a Wagner Safe-T-Planer, which is an inexpensive rotary planer that I chuck in my drill press.

The only drawback I encountered while rotary planing stock was the tendency for the wood to lift up, especially at the beginning and end of a pass, resulting in pieces that were unevenly thicknessed. To solve this problem, I built a vacuum hold-down table for my drill press, as shown in the photo at left. The hold-down surface's holes go through the tabletop and into a labyrinth (vacuum chamber), which is connected to an ordinary shop vacuum. The vacuum holds thin stock flat against the table, enabling me to plane pieces down to $1/32$ in. and up to a $1/2$-in. maximum thickness. Although I use my hold-down table for planing, I suspect that with a few modifications to clamp it to a benchtop, the table could be used for light routing and sanding.

Evan Kern built a vacuum table to hold down thin stock when he's rotary planing with his drill press. To increase the table's suction, he covers holes ahead of the cutter with cardboard. Here, Kern advances the cardboard with a walnut workpiece as he guides it along a fence that's clamped to the table.

Constructing the hold-down

The hold-down table consists of a 3/4-in.-thick medium-density fiberboard (MDF) tabletop mounted to a hollow base. The drawing at right shows the size and pattern of the holes to bore through the top. Three pieces of 3/4-in. plywood—the center one being the labyrinth—make up the base. A 1/2-in. plywood bottom is screwed to the base to provide ears for clamping. Two requirements that may be different for other drill-press tables and shop vacuums are the size of the bottom (mine is 15 in.) and the size of the vacuum opening (mine fits a 1 1/8-in.-OD PVC coupling).

Labyrinth

In addition to joining the holes in the table to the vacuum source, the labyrinth supports the workpiece beneath the cutter. The suction from even a small vacuum can distort the table if it's not adequately supported. After scroll sawing out a labyrinth (see the pattern in the drawing), cut out the other two base pieces and sandwich and glue the labyrinth between them. After the glue has dried, drill a hole for the vacuum hose.

Adding a control gate and a fence

If all the holes in the hold-down table are covered by a workpiece, there will be no relief for the vacuum and, as a result, your vacuum's motor may overheat. You can eliminate this problem by making a vacuum-control gate, which allows air to enter the labyrinth. I made a simple gate (see the photo on p. 150) out of 1/4-in. plywood. The gate slides over a pair of 1/2-in. holes bored in one side of the base. I can open the gate fully or partially to equalize the pressure in the labyrinth and regulate the degree of suction at the hold-down surface.

To guide stock when planing, I made a plywood fence that I spring clamp to the tabletop. The fence has a recess that lets the Safe-T-Planer overlap the edge of the work. I faced the underside of the fence with 1/32-in. plywood to cover the holes that would otherwise be exposed by the recess and to provide an edge for the workpiece to ride against at the recessed area.

Drill-press vacuum table assembly __

Hold-down table's base consists of three layers: upper, labyrinth and lower. Air flows into tabletop through upper and labyrinth, then out through vacuum hose. Tabletop holes are chamfered with countersink.

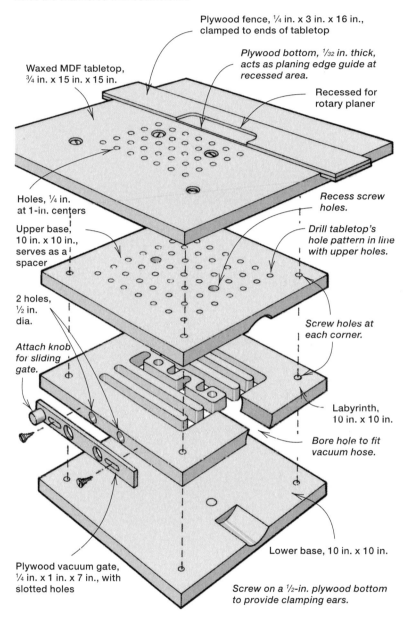

Plywood fence, 1/4 in. x 3 in. x 16 in., clamped to ends of tabletop

Plywood bottom, 1/32 in. thick, acts as planing edge guide at recessed area.

Waxed MDF tabletop, 3/4 in. x 15 in. x 15 in.

Recessed for rotary planer

Holes, 1/4 in. at 1-in. centers

Recess screw holes.

Drill tabletop's hole pattern in line with upper holes.

Upper base, 10 in. x 10 in., serves as a spacer

2 holes, 1/2 in. dia.

Attach knob for sliding gate.

Screw holes at each corner.

Labyrinth, 10 in. x 10 in.

Bore hole to fit vacuum hose.

Lower base, 10 in. x 10 in.

Plywood vacuum gate, 1/4 in. x 1 in. x 7 in., with slotted holes

Screw on a 1/2-in. plywood bottom to provide clamping ears.

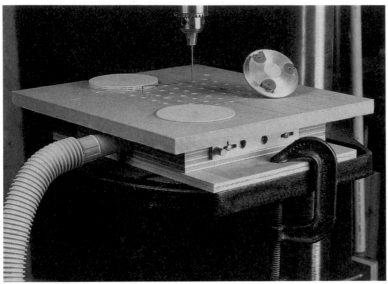

Before mounting his Safe-T-Planer, the author chucks in a bent piece of wire to level the hold-down table. While hand-turning the chuck, he observes and shims the table until the wire's tip grazes the surface for a full revolution. Then, using discs of modelmaker's plywood to gauge thickness, Kern will set the planer's depth of cut.

Rotary planing on the hold-down table

A Safe-T-Planer consists of a shaft connected to a 3-in.-dia. disc that holds three circular cutters (see the photo above). Because the cutters only project about $1/64$ in. from the disc, these rotary planers are quite safe. The planers, which will also work in most radial-arm saws, are manufactured by G & W Tool, Inc., PO Box 691464, Tulsa, Okla. 74169; (918) 486-2761 and are available at most woodworking supply stores. When using a rotary planer, the length of stock that can be planed is limited only by your shop space. The stock width is limited to your drill-press swing.

Squaring the table and setting the cut

To make sure the hold-down table is perpendicular to the drill-press spindle, I made a gauge by bending a heavy piece of wire into a Z-shape. After I mount the wire in the chuck, I rotate the chuck by hand and observe the gauge and the table's top. The gauge's tip should just touch the table's surface throughout its rotation.

I use 3-in.-dia. plywood discs as thickness gauges to set the height of the planer's cutter above my hold-down table. I bandsaw the discs from sheets of modelmaker's plywood (available at most hobby shops), which comes in precise thicknesses from $1/64$ in. to $1/2$ in., in $1/64$-in. increments. After placing a disc of the desired (planed) thickness on the table, I adjust the quill until the cutter just touches the gauge, and then I lock the quill.

Feed and cutter speed

After lightly waxing the tabletop, I hook up my shop-vacuum hose and turn my drill press on. If the wood is wider than the planer, after an initial pass, I reverse it end for end and continue passes, moving the fence in toward the drill-press column until I've planed the entire width of the board. I feed stock at a rate of approximately two to three square feet of wood per minute. At this rate, I've never had to sharpen the cutters, even though the manufacturer supplies instructions for this. Although the planer's maker recommends speeds of 3,000 RPM to 6,000 RPM, I've found that 2,300 RPM helps prevent the cutters from burning the wood during those inevitable feed pauses.

Regulating the suction

When I'm feeding narrow strips of wood into the planer, most of the hold-down table's holes are uncovered, and as a consequence, there's an insufficient vacuum. I resolve this by covering exposed holes with pieces of cardboard or stiff plastic. Feeding work against the planer cutters (from left to right) pushes the covers out of the way (see the photo on p. 148). As the end of a board is reached, I reintroduce another cover so that the holes in the table are continuously covered to maintain a vacuum.

COMPRESSION CHUCK FOR A LATHE

by Dale Ross

A nicely finished foot on the bottom of a turned bowl is one feature separating the work of a pro from that of a beginner. A well-proportioned foot lifts the bowl and gives it a classic look typical of pottery. Turning a foot also eliminates the mounting screw holes on the bottom of the bowl.

The biggest problem with creating a foot or finished bottom is not how to shape it, but how to hold the bowl in the lathe. Turning the foot is the last thing I do, so the outside and inside waste of the bowl has already been cut away and sanded, leaving no place for mounting screws. That's where my shop-built compression chuck comes in, making it possible to remount the bowl and complete the foot. The real advantage of this system is that once the chucks are made, they can be used over and over again. My set of four chucks will handle bowls ranging from 4 in. to 14 in. dia. The chucks are easy to make and inexpensive, too, because they're made from plywood and mahogany or poplar scraps.

How a compression chuck works

A compression chuck consists of a flexible jaw plate pressed to a curved baseplate by a platen, as shown in the drawing at right and the top photo on p. 152. A handwheel is screwed to the outboard end of a threaded rod that passes through the lathe's headstock. Tightening the handwheel draws the platen toward the headstock and squeezes the jaw plate between the baseplate and the platen, constricting the jaws of the chuck. As the jaws close in, they grab and hold the rim of a bowl.

Compression chuck _____

This shop-built lathe chuck grips a bowl firmly around its rim, so a foot can be turned. Because several steps are cut into the inside of the jaw plate, the chuck can handle bowls in a variety of sizes.

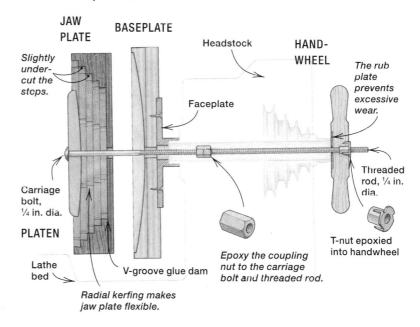

JAW PLATE BASEPLATE Headstock HANDWHEEL

Slightly undercut the steps.

Faceplate

The rub plate prevents excessive wear.

Carriage bolt, ¼ in. dia.

PLATEN

Threaded rod, ¼ in. dia.

T-nut epoxied into handwheel

Lathe bed

V-groove glue dam

Epoxy the coupling nut to the carriage bolt and threaded rod.

Radial kerfing makes jaw plate flexible.

HOW IT WORKS

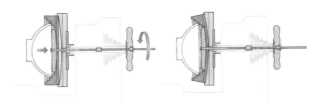

With the handwheel loosened, the bowl slides easily into the flexible jaw plate. As the handwheel is tightened, the jaw plate is compressed by a curved platen and captures the outer rim of the bowl.

The parts are simple. A compression chuck consists of a platen, jaw plate, baseplate and hand-wheel (from the left), all connected with a threaded rod. Tightening the hand-wheel flexes the jaw plate, so it grips the edges of a bowl.

A compression chuck holds a bowl securely at its rim while its foot is turned.

The jaw plate has a series of evenly stepped ridges to accommodate bowls of varying diameters. The compression chuck shown above is 11 in. dia. and will accommodate bowls from about 9 in. dia. to $10^5/8$ in. dia.

Making the baseplate

The baseplate is two pieces of plywood glued together, turned and hollowed out, as shown in the drawing. For an 11-in.-dia. chuck, glue and screw together two pieces of $3/4$-in.-thick by 12-in.-sq. plywood. Once the glue dries, remove the screws, mark the center and bandsaw the plywood to as large

a disc as possible. Temporarily mount the disc to a faceplate, and turn the outside edge true. Then cut a mortise into what will become the back side of the baseplate to match your faceplate (I used a 6-in.-dia. faceplate).

Better yet, leave a faceplate on each chuck. I make extra faceplates from 1-in.-thick aluminum plate, bandsawn round, drilled and threaded to my lathe shaft size. After screwing the aluminum faceplate onto the lathe shaft, I true it round and flat with high-speed steel tools.

To finish up the baseplate for the chuck, remove it from the lathe, and remount it on a faceplate screwed into the turned mortise. On the face of the baseplate, cut a shoulder, and then dish out the face of the plywood, as shown in the top photo at right. Go about ⁵⁄₈ in. deep, taking care not to hit the mounting screws. Try to achieve a nice, fair camber. Finally, drill a ¹⁄₂-in.-dia. hole through the center of the baseplate for the mounting bolt and threaded rod.

To help get the shape right, bandsaw a curved template out of ¹⁄₄-in.-thick plywood (I use a set of trammel points). The offcut will be the template for turning the platen, so hang on to it.

Turning the jaw plate

The jaw plate is the part that actually does the gripping. It's made of two pieces: a thin, flexible plywood backing and an outer ring of solid stock turned to form steps that grip the edge of a bowl. Evenly spaced sawkerfs around the perimeter of the jaw plate allow it to flex as it's squeezed between the platen and baseplate.

For the backing, use ¹⁄₈-in.-thick Baltic birch for chucks of 11 in. and less in diameter and ¹⁄₄-in.-thick Baltic birch for larger chucks. For the outer ring, glue up 8/4 poplar or mahogany into a 12-in. square, and bandsaw it round. After flattening the back of the solid disc and drilling a small hole through its center, I glue it to the plywood backing, but only around the perimeter. When I cut the final step in the blocking, the center section will fall away without a lot of unnecessary lathe work.

One thing that helps keep the center section from being glued to the plywood is a V-groove cut into the back of the disc that serves as a glue dam. The V-groove is cut just outside of where the last step will fall, as shown in the top drawing on p. 151. Apply glue only to the solid wood, outside the stop groove, and glue the solid-wood disc to a slightly larger plywood disc.

Drill a small center hole through the plywood using the previously drilled hole through the solid wood as a guide. This will locate the faceplate on the back side of the plywood. Mount the glued up disc on your lathe, and turn the outside diameter to match the inside diameter of the shoulder turned into the baseplate. Now turn the

The baseplate is turned from plywood. A template on a lathe bed helps the author shape a camber in the baseplate (colored yellow in the drawing on the p. 151).

A series of steps in the jaw plate of the compression chuck accommodate a range of bowl sizes (the jaw plate is red in the drawing on the p. 151).

steps to form the jaws into the face of the solid wood, taking care on the last step not to cut into the plywood (see the photos above). Slightly undercut the sides of each step for a better grip. I make the steps the same width as my parting tool (³⁄₈ in.), so I can cut each step quickly and accurately without measuring. The screws from the faceplate hold the unglued center area in place while turning. Once off the lathe, this center area of the hardwood disc should come right out.

Radial sawcuts, ¹⁄₄ in. wide (cut from the perimeter of the disc to within 3 in. of the center) divide the disc into eighths and allow the jaws to flex during compression (see the top photo on the facing page). If the jaws seem too stiff, make the radial cuts a little longer. A ¹⁄₂-in.-dia. center hole provides clearance for the threaded rod.

The platen and handwheel

The platen is turned from another piece of ³/₄-in.-thick Baltic-birch plywood. Mount a bandsawn, round piece of plywood on the lathe. Turn a crown into the face, matching the camber of the dished-out baseplate. Here's where the other half of the template comes in handy (see the photo below).

Drill a hole, and insert a ¹/₄-in.-dia. carriage bolt from the flat side of the platen. Attach a length of ¹/₄-in.-dia. threaded rod to the end of the bolt with a coupling nut, and then epoxy the joint. The bolt/rod combination should be long enough to pass through the platen, jaw plate, baseplate, lathe headstock and handwheel, as shown in the drawing.

The handwheel, which tightens the jaw plate around a bowl, is turned from hardwood. Epoxy a large washer to the inside face of the handwheel to act as a rub plate. This washer must have an inside-diameter hole large enough to allow the threaded rod to pass through it and an outside diameter large enough to cover the end of the lathe's spindle. Insert a T-nut into the outside face of the handwheel so that it can screw onto the threaded rod. Put the whole rig together on the lathe, and then hacksaw off any extra threaded rod. Leave enough of the threaded rod to engage the nut in the handwheel completely when the jaws are fully relaxed.

Using the compression chuck

With the chuck mounted on the lathe and the lathe's spindle locked, hold the bowl into the closest-fitting step of the chuck. For in-between sizes, I tape small pieces of ¹/₈-in.-thick plywood to each jaw of the next larger step with double-faced tape, but this is rarely necessary. Tighten the handwheel securely while holding the bowl solidly to the bottom of the step.

The closer a bowl's shape gets to perpendicular at the rim, the less secure the bowl is in the chuck. In this situation, I bring the tailstock up and sandwich the piece in with a long, blunt insert in the revolving center, allowing room for the tool-rest base, as shown in the bottom photo on p. 152. A center cone, which needs to be cleaned up with a sharp chisel, remains after turning. With light cuts and moderate spindle speed, I can turn a foot on a variety of bowl sizes without any problems.

The platen is turned with a crown to match the dish in the baseplate. The curved platen (green in the drawing on p. 151) flexes the jaw plate.

ADJUSTABLE LATHE JAWS

by Jim Leslie

A finely produced turning should display no telltale signs of how it was mounted on a lathe. Unlike marking-gauge lines on dovetailed furniture, grip marks on the base of a bowl are not meant to be seen. To get rid of them, you need to be able to mount your work so that it is held by its rim, and then turn the base. Large, adjustable jaws with rubber bumpers are commercially available, but they cost about $100. After looking them over, I decided I could build my own with scrap plywood and a few dollars worth of materials. It took me about three hours to construct my adjustable jaws, and after countless hours of use, I'm very satisfied with the results.

I built my jaws to fit an adjustable, four-jaw Nova chuck, but the fixture may be adapted to other four-jaw chucks by using the same procedures. These jaws allow me to mount bowls, rings and plates of different sizes. The jaws will also hold oddly shaped pieces.

Make the body of the jaws out of plywood

Every woodworker I know has a few scraps of plywood lying around. A 12-in. square is about all you need. It should be free of voids and at least $1/2$ in. thick. The diameter of your jaws will depend on the swing of your lathe. Make them about $1^1/2$ in. to 2 in. less than the maximum swing so that when you fully open your chuck, the jaws don't strike the lathe bed.

Using a straightedge and a sharp knife, connect the corners of the plywood square, marking two diagonal lines to divide it into quarters. Where the lines intersect, place a compass point, and draw a 12-in.-dia. circle.

Cut the piece into quarters. (If necessary, plane or resaw any pieces to get them all the same size.) Then cut the curved sections on the bandsaw.

Place the pieces together to form a circle, and use your four-jaw chuck as a guide to mark the mounting holes. It's important that each quadrant attaches to your chuck with at least two fasteners. Drill and countersink each hole. Next mark the locations for the adjustable stops. Using a protractor on one of the segments, divide it into four equal parts, and draw lines for the three bisecting angles (see the drawing on p. 156). On one of those lines, place seven tick marks spaced $1/2$-in. apart. Mount the pieces

Shopmade adjustable jaws—Eight stops securely hold a workpiece for turning.

Shopmade adjustable jaws

These jaws will securely hold a turning by its rim, allowing you to finish working on the base. Size the plywood so that the finished assembly is 1½ in. to 2 in. less in diameter than the swing of the lathe.

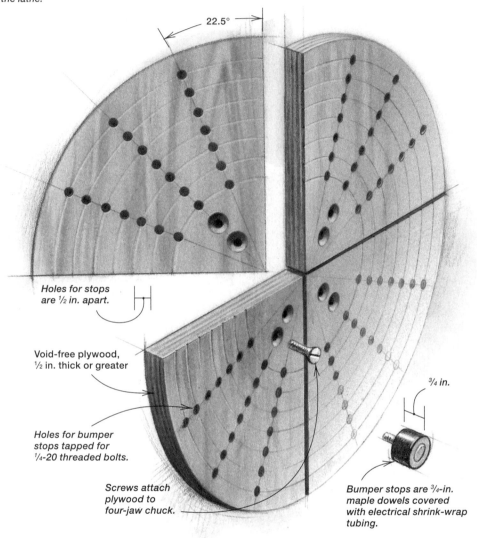

22.5°

Holes for stops
are ½ in. apart.

Void-free plywood,
½ in. thick or greater

Holes for bumper
stops tapped for
¼-20 threaded bolts.

Screws attach
plywood to
four-jaw chuck.

¾ in.

Bumper stops are ¾-in.
maple dowels covered
with electrical shrink-wrap
tubing.

on the chuck, and with the lathe turning at a slow speed, touch a skew to the ½-in. marks. Drill the holes for the stops where the skew marks intersect the radial lines. Before removing the segments for drilling, take a gouge or scraper to the outside edge of the plywood and turn it round.

After detaching the plywood segments from the chuck, stack and clamp them on the drill-press table, and bore the holes for the stops using a $^{13}/_{64}$-in. drill. Finally, tap all the holes with a ¼-in.-20 tap using a reversible drill. If you plan to turn irregularly shaped pieces, you can machine a long, ¼-in.-wide slot at the 45° marks, which will give you infinitely variable attachment points for four bumpers. (Use longer bolts for those bumpers, and secure them with nuts.)

Screws hold jaws to a lathe's chuck. The four-piece jaws can be attached to any four-jaw lathe chuck.

Use straight dowels for the bumpers

Dowel stock is sometimes more oval than round. Select round stock so that your bumper stops will exert even pressure on workpieces. I used ³/₄-in. maple dowels, which I cut into ³/₄-in.-long segments. Each of the eight dowel pieces was countersunk ¹/₈ in. deep exactly in the center with a ¹/₂-in.-dia. brad-point drill. This is best done on the lathe so everything is centered. After countersinking each piece, I switched to a ¹/₄-in. drill and bored all the way through each one.

I fit 1³/₈-in.-long, ¹/₄-20 bolts through each piece of dowel, recessing the head into the countersunk hole, and glued them in place with epoxy. Then I encased each dowel with a piece of shrink-wrap electrical tubing, which helps grip the workpiece without marring it.

Give the jaws a test run

Assemble the segments on the chuck, and test the operation. You should be able to expand and contract your chuck without the jaws binding. They should also nest together flat when fully contracted. Mount something to the jaws that you know is round, such as a pie plate, to see whether the stops are concentric. If some of the stops don't close in properly, start over using a new piece of plywood.

Turning a base while leaving no marks—this shopmade jig gets you there at a fraction of a commercial jig's cost.

INFEED/OUTFEED TABLE FOR A PORTABLE PLANER

by Greg Colegrove

Portable thickness planers that I've used have an annoying habit of sniping the ends of boards. My 12-in. Delta planer is no exception. I initially accepted that I'd have to scrap $2^1/_2$ in. on both ends of every board I planed. But soon my conscience, spelled checkbook, convinced me there had to be a better way. Because I really am a rocket scientist, I figured I should be able to cure this otherwise fine machine of its hiccups.

Snipe is a deep cut, like a divot, in one or both ends of a planed board (see the inset photo on the facing page). Snipe occurs when the end of the board tilts upward into the cutterhead. Portable planers are known to be snipers because of their short, roller-less beds. Planers that are adjusted for depth of cut by raising and lowering the head are particularly susceptible.

This problem is hard to correct on many small planers because they don't have large, stable beds. A machinery engineer I spoke with confided that the best I could expect with a planer like mine, without any modifications, was about 0.005 in. of snipe. I knew I could do better, so I started looking for ways to improve support for the workpiece as it passed through the planer.

Designing a rigid, height-adjustable auxiliary bed

My first effort consisted of extension rollers, which required far too much fiddling, and the rollers had no guides to keep stock mov-ing straight. Then I saw a planer auxiliary bed at a local woodshop that went right through the mouth of the planer. I went back to my shop and adapted the idea, making a table that was stiff but adjustable in height (see the photo on the facing page).

The adjustable bed has reduced my planer's snipe to between 0.002 in. and 0.003 in. Such a small discrepancy in thickness is easy to sand or handplane out. And with the long bed, I don't have to feed and retrieve one board at a time. Stock up to about $4^1/_2$ ft. long is supported at the far end, so I can plane several in a bunch, left to right, or in succession, end to end. Gang-feeding eliminates snipe altogether.

The auxiliary bed reduces my planer's depth-of-cut capacity by $^3/_4$ in., but I rarely plane anything thicker than 4 in. anyway. If I'm planing large, heavy planks, I put blocks under both ends of the table to avoid damaging or moving them.

Choose from two different support platforms

I've built two different versions of the auxiliary bed: one in which the feed table sits on a dedicated base and one that can be moved. The only difference between the feed tables is length. The first bed, which sits on a cabinet platform, is designed for stock up to 6 ft. long.

DANGER CUTTERHEAD KEEP HANDS AWAY

A flat, stable planer bed reduces snipe. The author's height-adjustable feed table and platform, both made of melamine, have practically eliminated the problem. Maple edging prevents boards from wandering off the table.

This planer table can be moved. A feed table attached to a portable base allows a planer to be parked in a corner when it's not needed. The base can be set up quickly on any stable surface, like a table or a benchtop.

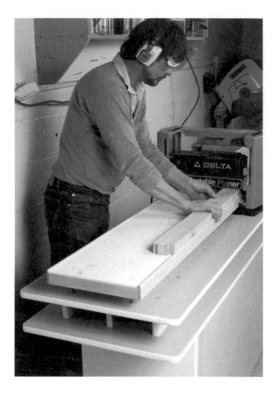

If space is a problem, build a portable base

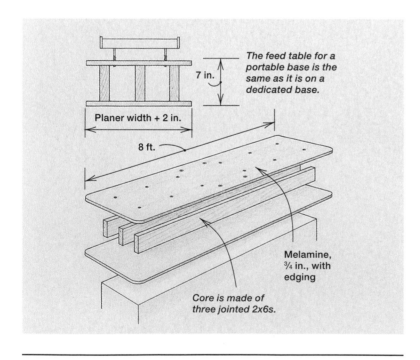

The feed table for a portable base is the same as it is on a dedicated base.

7 in.

Planer width + 2 in.

8 ft.

Melamine, ¾ in., with edging

Core is made of three jointed 2x6s.

The second bed, which I built for a cabinetmaker friend, is 8 ft. long and has a more compact platform. It must be placed on a stable bench or work table (see the photo at left), but it will handle longer stock and is more portable. With the help of a buddy, you could stand the unit—with the planer attached—against a wall, store it overhead or transport it in a truck or van. Most portable planers don't come with stands, so it's nice to have the planer at a comfortable operating height.

Inexpensive materials and hardware

You should be able to build either one of these planer auxiliary beds for less than $75. I made both units primarily out of white melamine, which provides a flat, low-friction surface for the feed table. And with melamine, it's easy to spot and remove wood chips and debris. For a rigid table, I fastened two layers of ¾-in. melamine together.

You will need a bit of hardwood to make edge guides for the tables. I used ¼-in.-thick maple strips that extend ¼ in. above the surface. I also covered the melamine ends with maple trim, this time flush with the tabletop. Not too long ago, I added a planer hood that's connected to a dust collector. The table should be kept free of chips; otherwise, boards will plane inconsistently.

I can adjust the flatness of the feed table with 12 connector bolts that join the table to its support platform. The heads of the bolts are sandwiched between the table's two layers of melamine. The bolts screw into threaded inserts set in the top of the platform (see the inset drawing on the facing page). The connector bolts and threaded inserts can be purchased from The Woodworkers' Store (4365 Willow Dr., Medina, MN 55340; 800-279-4441).

Building the feed table and platform

You can construct the auxiliary bed and platform in about an afternoon. You will need three pieces of melamine for the feed table. Two of the pieces—one for the outfeed side and one for the infeed side—are identical. The third is the continuous top of

Auxiliary bed construction

The bed consists of a feed table and a platform. Connector bolts adjust height of infeed and outfeed ends of table. Edge guides keep boards straight during planing. Platform can be either fixed or portable (see the drawing on the facing page). Modify dimensions to suit your portable planer.

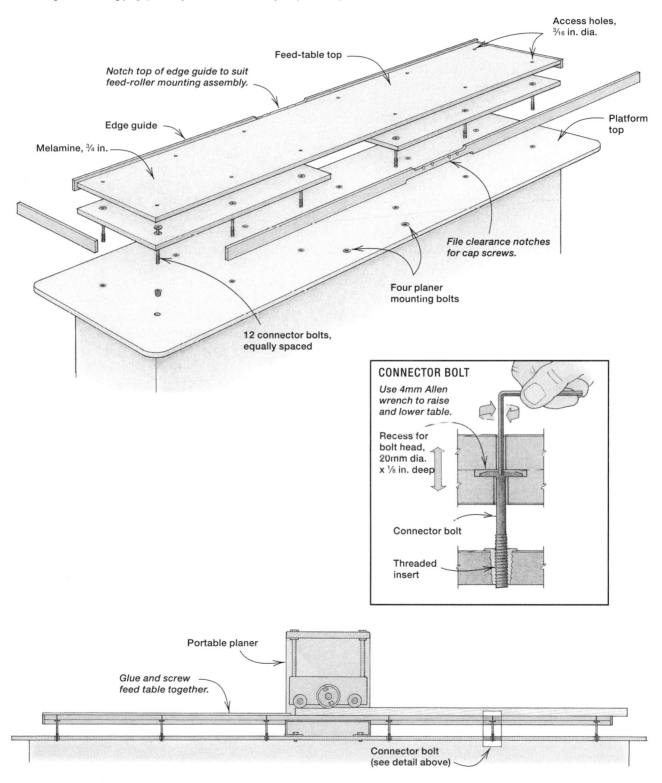

Access holes,
³⁄₁₆ in. dia.

Feed-table top

Notch top of edge guide to suit feed-roller mounting assembly.

Edge guide

Melamine, ¾ in.

Platform top

File clearance notches for cap screws.

Four planer mounting bolts

12 connector bolts, equally spaced

CONNECTOR BOLT

Use 4mm Allen wrench to raise and lower table.

Recess for bolt head, 20mm dia. x ⅛ in. deep

Connector bolt

Threaded insert

Portable planer

Glue and screw feed table together.

Connector bolt (see detail above)

the table (see the top drawing on p. 161). Passing only one thickness of melamine through the planer minimizes the loss of cutting depth. This allows enough table flexibility for independent height adjustment at the infeed and outfeed ends.

The bottom two pieces are placed just up to the front and back of the planer base. When you have all the table pieces, cut out a piece of melamine for the top of the platform, and place it on a long workbench. Stack the three table pieces on top of the platform top in the right order and relative positions. Temporarily align and clamp the pieces together.

Lay out the six rows of connector bolt holes, mark them with a center punch and drill $^3/_{32}$-in.-dia. pilot holes through the stack. Unclamp the pile, and remove the piece that will become the feed-table top. Counterbore clearance holes for the heads of the bolts in the two halves of the lower table layer. I used a 20mm Forstner bit (about $^{13}/_{16}$ in.). Bore these holes $^1/_8$ in. deep, slightly more than the thickness of a bolt head. Drilling any deeper will only increase backlash and make table adjustment more difficult. Use a $^{17}/_{64}$-in.-dia. bit to enlarge the pilot holes in both lower table sections. These holes should allow a connector bolt to fit easily through. Move back to the feed-table top, and using a $^3/_{16}$-in.-dia. bit, ream out the pilot holes. These will be

the access holes for raising and lowering the table with a 4mm Allen wrench (see the inset drawing on p. 161).

Assemble the parts in stages

On another flat surface in the shop, insert all 12 connector bolts between the two table layers (three pieces total). Glue the layers together with construction adhesive; keep the glue away from the bolt heads. Clamp the assembly, place it on its back and then fasten the layers together using 11/4-in. #6 particleboard screws, placing them about 6 in. on center. Let the inverted table cure flat.

Move back to the bench so you can finish the platform. Bore out the pilot holes in the top, and install 12 threaded inserts. To prevent them from vibrating loose, I epoxied them in. Cut out and assemble the pieces for the rest of the platform style you've selected, and attach the platform top.

Finally, rip some hardwood strips for the table's edge guides and end caps. Notch the bottom center of the guides to clear the base of the planer. I wanted to be able to plane $^1/_8$-in.-thick stock, so I also trimmed the top of the guides to fit under my Delta 22-540's feed-roller supports, and I filed clearance notches for their cap screws. Glue and clamp the edge guides and end caps to the feed table.

Mounting the planer and the table to the platform

Place the planer on the center of the platform, and transfer the bolt-hole locations (one at each corner) from the planer's base. Remove the planer, and drill holes for threaded inserts at those spots. I used $^1/_4$-in. bolts, nuts and washers to mount mine. If your planer has bed-extension wings or other parts that might interfere with the feed-table installation, remove them.

Leveling the table—Connector bolts screwed into threaded inserts allow precise adjustments to the height of the planer table.

Before you install the feed table, it's a good idea to thread a ¼-20 nut on each connector bolt. The nuts will keep the bolts from vibrating loose and take up play in the threaded inserts once you've trued up the table. Crank the cutterhead all the way up, slide the feed table through the planer and lower the table until the connector bolts align with the threaded inserts. As you snug the bolts, you'll find that you can only tighten each bolt a few turns before the table starts to bind. Move on to an adjacent bolt, and then continue around the table. Repeat this sequence until the table's edge guides are nesting over and almost touching the base of the planer.

Reset the depth scale—To correct a ¾-in. loss of cutting depth, install a new scale or reset the old one.

Fixing the platform and adjusting the tables

It's now time to adjust the table for flatness. Place the platform, with planer, where you want it. If it's the portable platform, make sure it's secured to a bench or table. Check that the space between the top of the planer base and the bottom of the feed table is free of sawdust. Tighten the two connector bolts on each side of the planer until the table touches the base. Don't overtighten these bolts. Hold a long (at least 5 ft.) straightedge on the infeed end of the table, and starting at the next pair of bolts, tighten them until that table section is flat (see the photo on the facing page).

Switch to the matching set of bolts on the outfeed end, and do the same procedure. Work outward and back and forth until the entire table is flat. If you installed nuts, snug them to the platform. To reduce friction, I periodically spray the feed table with TopCote, a dry lubricant.

Checking planer and depth-of-cut settings

You should check that the planer's two feed rollers are at the right height in relation to the cutterhead and that the knives and rollers are parallel to the table (for more on making planer adjustments, see *FWW* #107, pp. 72-77). The last step before using your auxiliary bed is to reset or replace the thickness indicator to account for the ¾ in. loss in depth of cut. You can recalibrate the original gauge or apply a new scale on the planer, offsetting it by ¾ in.

An accurate way of calibrating the scale is to plane a piece of scrap to a known thickness, and align the scale so that the indicator points to that measurement (see the photo above).

PORTABLE-PLANER CHIP COLLECTOR

by George M. Fulton

The first time I used my new portable planer, I realized that it needed a chip-collection system. As it was, shavings and dust were streaming out of the discharge chute, floating in the air and settling on my work, the table, the floor and on Goldie, my yellow Labrador retriever, who sleeps nearby.

I decided to make a chip-collector cabinet that would remove dust and serve as the base for the machine (see the photo below). The cabinet had to be compact, connect easily to my Delta planer without substantial modification, and it had to be inexpensive and easy to build. It also had to be stable, like a stand, but mobile so I could wheel the planer out of the way.

Planer chips conquered—To tame his biggest chip maker, George Fulton took a discarded vacuum motor and built this combination planer stand and dust cabinet. By attaching a wand to the flexible hose, he can vacuum up leftover dust.

Construction

You should be able to adapt the cabinet to any machine by slightly modifying the dimensions or construction shown in the drawing on the facing page. Basically, the cabinet consists of a frame boxed with plywood, a vacuum compartment and top made of plywood, a discarded vacuum motor and a plastic cat litter tray. On the infeed side of the cabinet, I mounted a screened vent to cool the motor, and I built a drawer to hold miscellaneous adapters and tools. On the side of the cabinet, I made a vacuum inlet, and on the outfeed end, I added a clean-out door. I fashioned a dust hood (the manufacturer didn't offer one at the time) out of sheet metal, which mounts to the planer and provides a way to connect the dust-inlet hose to the cabinet.

Dust hood

I formed the dust hood out of 0.017-in.-thick (27 gage) galvanized sheet metal, making sure that the hood wouldn't interfere with the chip deflector or a workpiece. I riveted the hood to the guard, and then I installed 2-in.-dia. flexible hose, which was compatible with the PVC pipe fittings I had. If you want to hook the cabinet up to a standard collector, you'll probably want to use 3-in. or 4-in. hose and fittings. Because the planer's thicknessing range is achieved by raising and lowering the cutterhead, I used flexible hose. That also lets me easily disconnect it from the hood, pop on a standard vacuum pick-up wand and clean up dust around the planer.

Planer cabinet

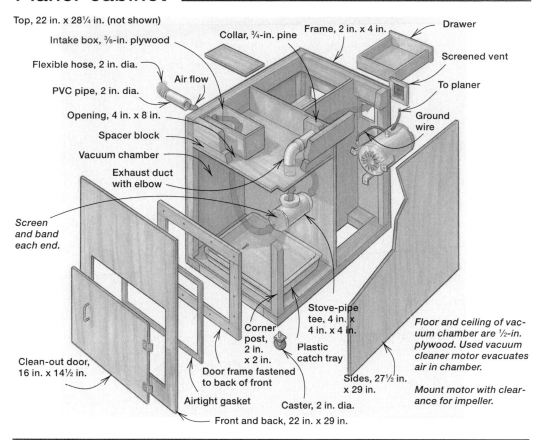

Top, 22 in. x 28¼ in. (not shown)

Intake box, ⅜-in. plywood

Flexible hose, 2 in. dia.

PVC pipe, 2 in. dia.

Air flow

Opening, 4 in. x 8 in.

Spacer block

Vacuum chamber

Exhaust duct with elbow

Collar, ¾-in. pine

Frame, 2 in. x 4 in.

Drawer

Screened vent

To planer

Ground wire

Screen and band each end.

Clean-out door, 16 in. x 14½ in.

Corner post, 2 in. x 2 in.

Door frame fastened to back of front

Airtight gasket

Front and back, 22 in. x 29 in.

Plastic catch tray

Caster, 2 in. dia.

Stove-pipe tee, 4 in. x 4 in. x 4 in.

Sides, 27½ in. x 29 in.

Floor and ceiling of vacuum chamber are ½-in. plywood. Used vacuum cleaner motor evacuates air in chamber.

Mount motor with clearance for impeller.

Cabinet

I constructed the frame, as shown in the drawing above, using 2x4s and 2x2s. For the box sides, top and vacuum chamber, I used ½-in. and ⅜-in. plywood. To make clean-up easier, I adhered plastic laminate to the cabinet top. I mounted a discarded vacuum-cleaner motor in the exhaust chamber using aluminum angle brackets. I grounded the motor housing using a lug terminal and the green wire of the motor cable. After I installed the vent and drawer on the infeed end, I added a partition between them, so the air is exhausted through the vent and the opening in the floor of the box.

Although the vacuum compartment's seams were tight, I applied a bead of caulk all around the interior corners and inlet box joints. Because the clean-out door had potential to leak air, I surrounded the opening's inner frame with a ¼-in. by ¾-in. weather-seal gasket. A pivoting latch compresses door to gasket.

Duct work

The exhaust duct consists of a pine collar, 2-in. PVC pipe and elbow, and a 4-in. tee. I formed a section of aluminum window screen over the two open ends of the tee and secured them with rubber bands. You could cover the screen with nylon stocking to further filter dust. For the inlet duct, I used 2-in. PVC, flexible hose and threaded coupling (see the photo on the facing page).

Wiring and final details

To allow the vacuum to run after the planer is off, I installed a toggle switch next to the cutterhead switch and wired it to the motor. After I secured four furniture casters to the bottom frame of the cabinet, I mounted the planer to the cabinet with bolts and T-nuts. Finally, I placed a plastic waste tray in the vacuum chamber under the inlet box to gather the lion's share of shavings.

Now my dog Goldie dozes fairly contentedly, although she is probably wondering if something can be done about all the noise.

ROLLING CHOP-SAW STAND SAVES SPACE

by Charles Jacoby

My shop is pretty crowded, so when I acquire a new tool, I have to create efficient ways to store and use the tool. Such was the case after I bought a new sliding compound-miter saw. The saw needed a permanent, but mobile, home where I could do accurate cutoff and miter work. I first tried using the saw on planks and horses. This worked fine for single cuts, but I really needed a fence with a stop for cutting multiples. And the extensions that came with the saw limited its cutting to short pieces.

After a well-executed cutting performance, Jacoby's cutoff-saw stand (with a sander stowaway) gets its wings lowered and is rolled to a tidy corner in the shop.

Also, I still had to break things down to put the saw away.

About this time, my wife, Rosemary, gave me a benchtop oscillating-spindle sander. Again I wondered where I would store the tool. Building a stand to house both tools was the answer—make that a movable stand with folding extension wings. I designed the stand with crosscutting and mitering in mind but with a place to store the sander. I also left room for a top drawer to hold my shaper cutters and accessories. When I'm not using the saw, I drop the wings and roll the stand into a corner (see the photo at left). And even with the wings folded down, I can still do short chop-saw work by clamping a stop block to the saw's auxiliary fence.

Cabinet construction

For the stand's carcase, I made a $^3/_4$-in. birch-plywood box. To make storing the sander easy, I left the stand's lower compartment open (back and front). I dadoed the box's top, middle and bottom $^1/_4$ in. into the sides. Then, using #10 biscuits and glue, I plate-joined maple face frames to the front and back of the carcase to make the box rigid. Because my miter saw has its own base with four feet, I recessed the top of the cabinet so that the saw's work surface would be at the same height as the wings (see the drawing on the facing page). I also fastened four $2^1/_2$-in.-dia. casters (two of them locking) to hardwood plates that I glued to the bottom of the cabinet. The added height of the casters puts the top of the stand at a

Anatomy of a mobile chop-saw stand

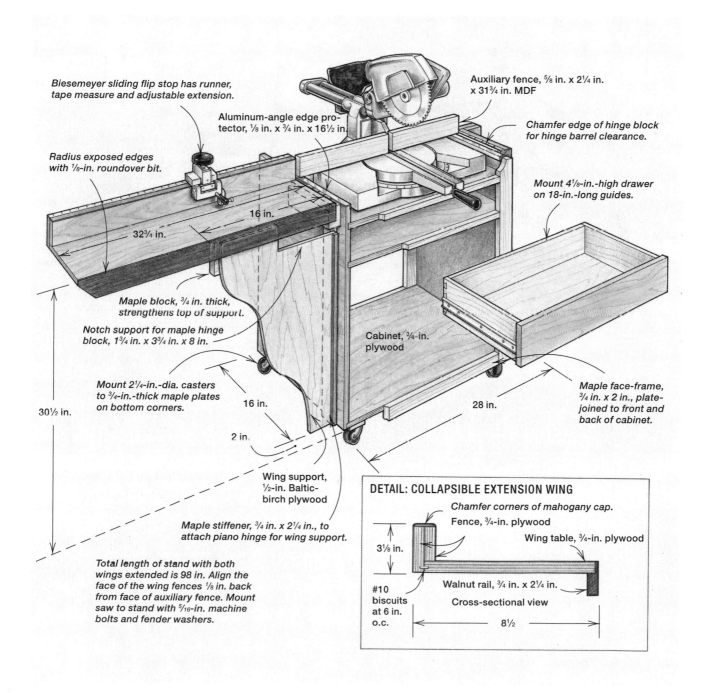

Biesemeyer sliding flip stop has runner, tape measure and adjustable extension.

Aluminum-angle edge protector, ⅛ in. x ¾ in. x 16½ in.

Radius exposed edges with ⅛-in. roundover bit.

Auxiliary fence, ⅝ in. x 2¼ in. x 31¾ in. MDF

Chamfer edge of hinge block for hinge barrel clearance.

Mount 4⅛-in.-high drawer on 18-in.-long guides.

16 in.

32¾ in.

Maple block, ¾ in. thick, strengthens top of support.

Notch support for maple hinge block, 1¾ in. x 3¾ in. x 8 in.

Cabinet, ¾-in. plywood

Mount 2¼-in.-dia. casters to ¾-in.-thick maple plates on bottom corners.

30½ in.

16 in.

2 in.

Maple face-frame, ¾ in. x 2 in., plate-joined to front and back of cabinet.

28 in.

Wing support, ½-in. Baltic-birch plywood

Maple stiffener, ¾ in. x 2¼ in., to attach piano hinge for wing support.

Total length of stand with both wings extended is 98 in. Align the face of the wing fences ⅛ in. back from face of auxiliary fence. Mount saw to stand with ⁵⁄₁₆-in. machine bolts and fender washers.

DETAIL: COLLAPSIBLE EXTENSION WING

Chamfer corners of mahogany cap.

Fence, ¾-in. plywood

Wing table, ¾-in. plywood

3⅛ in.

#10 biscuits at 6 in. o.c.

Walnut rail, ¾ in. x 2¼ in.

Cross-sectional view

8½

comfortable working level. To protect the top edges of the plywood sides, I mounted strips of $^3/_4$-in. by $^3/_4$-in. aluminum angle.

Collapsible wings

What makes the stand accurate and maneuverable are the folding pair of wings attached to the side of the cabinet. Each wing basically consists of a table, a support and a fence. The tables are $^3/_4$-in. plywood and the supports are made from $^1/_2$-in.-thick Baltic-birch plywood for strength. To stiffen the wing tables, I made front rails using walnut I had on hand. The wave-like curves of the supports aren't necessary, but I wanted to use my new spindle-sander. To strengthen the top of the supports, I glued and screwed on a maple block to the back side of each. Finally, I made the fence for each wing from two pieces of $^3/_4$-in. plywood, staggered and glued together to form a rabbet (see the drawing detail). I glued and biscuited the fences' rabbets to the wing tables, and then I capped the top of the fences with mahogany. I chamfered the caps' edges, so there would be enough clearance for the runner block of an adjustable stop.

A flip stop for the fence

By securing a flip stop to the left fence, I'm able to measure precise lengths. The stop I use is made by Biesemeyer Manufacturing Corp. (216 S. Alma School Rd., Suite #3, Mesa, Ariz., 85210; 800-782-1831). I purposely made my fence higher than what the flip stop requires to permit a full 2x4 to go under the stop. Because of the extra height, I had to make a metal stop extension to get it low enough for thin boards.

Aligning the wings and mounting the saw

The collapsible wings are strong; I can crosscut 14-ft.-long 2x8s in half on a fully extended stand. To achieve this kind of load, I had to first add blocks and stiffeners to reinforce the cabinet where the wing-table and wing-support hinges attach. I secured

$1^3/_4$-in.-thick support blocks to the top of the cabinet sides. Then I fastened $^3/_4$-in.-thick strips of maple to the plywood sides. For the hinges, I fastened two Corbin ball-bearing (large door) hinges to the wing tables and mounted a pair of 2-in. by 24-in. piano hinges to the wing supports.

Before I screwed the hinges to the cabinet, I lined up the tables and fences as follows: First, I propped each wing assembly in place with buckets and blocks. Next, I set my saw down at the rear of the cabinet top and laid a 6-ft.-long straightedge across the front of the fences. After I had shimmed each wing so its fence was properly aligned (an extra pair of hands are a big help), I flipped the straightedge 90° to set the height of the wing tables. Once the wings were in position, I carefully clamped the hinges in place, so I could make pilot holes. Finally, I screwed the table hinges to the support blocks and the piano hinges to the stiffener strips.

I offset the saw's auxiliary fence about $^1/_8$ in. ahead of the wing fences so that they won't influence the alignment of long boards held snugly to the saw's fence. I fastened the saw to the cabinet top using $^5/_{16}$-in. machine bolts with large fender washers under the plywood. With the wings extended, I originally figured I'd have to clamp the supports to the front rails. But the wing tables are heavy and rest on the supports unaided.

Finishing touches

To complete the stand, I made a simple drawer for the upper cabinet opening. Before installing the drawer on a pair of 18-in.-long slides, I notched the top of the drawer back to clear the ends of the saw-mounting bolts. Finally, I sealed the drawer, cabinet and wings with clear Watco oil. Once my mobile stand was finished, I put the saw right to work, cutting everything from baseboard to pull-out dish racks for the kitchen.

A STOP FOR EVERY JIG

by Sandor Nagyszalanczy

Most things that we do in our everyday lives have limits: the maximum speed you're supposed to travel on the highway; the minimum age you must be to buy a bottle of liquor; the most books you can check out of a library at one time. The world of woodworking is no different, except we call the limits *measurements*. We strive to maintain the exactness of measurements to make parts fit more precisely together, so the joinery will be strong and look clean. Some measurements are set on our machines, such as the depth of cut of a tablesaw or handplane, and some must be regulated by eye, as when chiseling down to a pencil line. But we regulate a great many limits—measurements for the length or width of parts, depth of grooves and holes—by using stops on our jigs and in conjunction with our tools.

Regulating the distance between the end of a part and the point where it's cut to length or machined is a basic function of stop devices. As with other types of jigs and shopmade setups, there are many different kinds of stops to choose from, each appropriate for a particular range of tools and applications. The simplest stops are merely wooden blocks, clamped or screwed to the machine, jig or the work itself. More ingenious stops revolve to adjust or change position. The right stop can increase the accuracy of an operation, as well as save time when making repeat cuts because parts need not be marked individually. This is why production shops can't do without the use of stops.

Basic flip-down stop

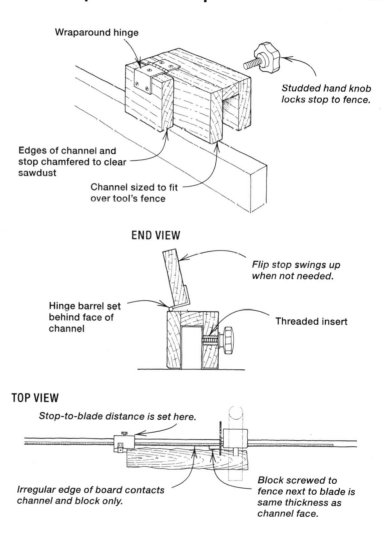

Wraparound hinge

Studded hand knob locks stop to fence.

Edges of channel and stop chamfered to clear sawdust

Channel sized to fit over tool's fence

END VIEW

Flip stop swings up when not needed.

Hinge barrel set behind face of channel

Threaded insert

TOP VIEW

Stop-to-blade distance is set here.

Irregular edge of board contacts channel and block only.

Block screwed to fence next to blade is same thickness as channel face.

Length stops

Length stops are used mostly for crosscutting or shaping across the width of stock, but they are easily adapted to work with other machines in a variety of applications. Length stops are commonly used on tablesaws, radial-arm saws, sliding-compound-miter saws and both powered miter saws and nonpowered (handsaw) miter boxes. Length stops are also welcome additions to fences used with miter gauges, drill presses, mortising machines, sliding crosscut boxes and other sliding carriage jigs.

While the stops described here are shopbuilt, there are several high-quality, commercially produced stop devices on the market, such as the FastTrack stop system components including the micro-adjusting FastStop (available from Garrett Wade, 161 Avenue of the Americas, New York, N.Y. 10013; 800-221-2942). Also, the ProScale digital readout (available from Accurate Technologies, 11533 N.E. 118th St., Suite 220, Kirkland, Wash. 98034; 800-233-0580) can be added to many of the shopmade stops described below.

Adjustable flip-down stops

Probably the most useful kinds of stops for basic crosscutting applications are adjustable flip-down stops. A flip-down stop is more useful than a simple stop block clamped to the fence because it quickly flips out of the way when it's not needed. This allows one end of the workpiece (a frame member or molding) to be squared with the stop flipped up. The part is then rotated end for end, and the stop (set and locked in the desired location) is flipped down to cut the part to final length. The two basic types of flip-down stops presented are illustrated as applied to a radial-arm or other crosscutting saw; however, they can be used as adjustable length stops on many other machines as well.

Setting a crosscut is a breeze with a flip-down stop. The cursor's cross hair of the T-track-mounted flip stop lines up with the desired measurement on a self-adhesive measuring tape stuck on the fence.

Track-mounted flip-down stop

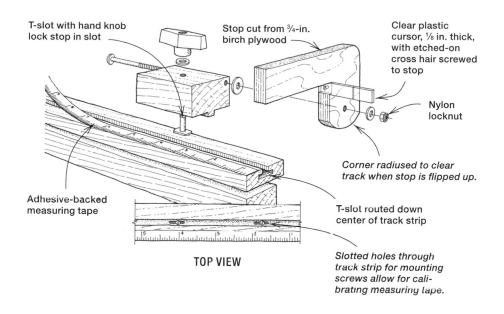

T-slot with hand knob lock stop in slot

Stop cut from ¾-in. birch plywood

Clear plastic cursor, ⅛ in. thick, with etched-on cross hair screwed to stop

Nylon locknut

Corner radiused to clear track when stop is flipped up.

Adhesive-backed measuring tape

T-slot routed down center of track strip

TOP VIEW

Slotted holes through track strip for mounting screws allow for calibrating measuring tape.

Cursors and stick-on metal rules improve accuracy

Thin metal rules with a pressure-sensitive peel-and-stick backing provide a convenient way to add an adjustment scale to any fence or adjustable jig component. Scales are available that read both right to left and left to right (available from Highland Hardware; 800-241-6748). Reading the position of the movable part can be done by simply mounting the scale underneath the part or by adding a fine cross-hair cursor to the moving part.

To make a cursor, start with a piece of clear plastic. Make a test cut with the cursor installed on the jig to determine the cross hair's exact location. Then etch the cross hair on the down-facing side of the plastic using a scratch awl and a try square (see the photo at right). Color in the cross hair with a thin-point permanent marker pen, applied judiciously, to make it easier to see. If you're using a stop fitted with a cursor on a radial-arm saw that uses dado blades or sawblades of various thicknesses, you can etch additional cross hairs on the cursor; position them so they will represent the location of the cuts produced by those blades. —*S.N.*

Etched cursors are easy to make. A thin line etched with a scratch awl onto a piece of clear plastic makes the cross hair for a cursor that mounts to a flip stop used on a cutoff saw. Permanent marker on the etched line makes it easier to see.

Basic flip-down stop

The flip-down stop shown in the drawing on p. 169 will work with just about any wood or metal crosscutting fence, and the stop can be set to any measurement, limited only by the length of the machine's fence. The channel-shaped body of the stop should be about 6 in. to 8 in. long and sized to fit not too snugly over the fence. A threaded insert driven into the back of the channel takes a studded hand knob, which locks the stop to the fence. The flip stop itself attaches to the channel with a wrap-around-style cabinet hinge, located so the hinge barrel is behind the front face of the channel (see the end view in the drawing on p. 169). This keeps the flip stop completely out of the way when it's up. The edge of the channel face and corner of the stop are chamfered to keep sawdust from misaligning the workpiece.

In use, the stock to be cut doesn't contact the machine's fence; one end bears against the face of the channel while the other bears

Flip stop for mitered ends

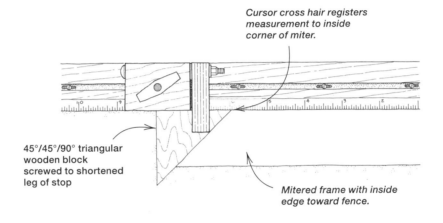

Cursor cross hair registers measurement to inside corner of miter.

45°/45°/90° triangular wooden block screwed to shortened leg of stop

Mitered frame with inside edge toward fence.

Measuring miters is easy with a dedicated stop. This flip stop has been fitted with a 45° block for mitered ends. The stop's cursor shows the distance between the inside corner of the miter and the miter created when the member is cut.

on a short block the same thickness as the channel that is screwed to the fence next to the blade (see the top view in the drawing on p. 169).

This arrangement allows you to cut stock that's bowed and won't set stably against the straight fence. The block near the blade also supports the workpiece near the cut to prevent tearout. To use this stop with a stick-on measuring tape, offset the tape's position, so the blade-to-stop distance can be set by aligning the end of the channel with the desired measurement.

Track-mounted flip-down stop

Another flip-down stop, as shown in the photo on p. 170, rides on and locks to a track strip. As shown in the drawing on p. 171, this adjustable stop setup has four basic pieces: a track strip with measuring tape, a sliding block, an L-shaped stop, and a cross hair and a cursor that allow very accurate settings. The solid-wood track strip has a T-slot routed in the top edge and an adhesive-backed, stick-on measuring tape pressed on. Flat-head screws through slots routed in the center of the T-slot mount the track to the top of the tool's fence. These slots allow side-to-side adjustment for calibrating the strip's measuring tape to the blade.

The sliding block has a short tongue that loosely fits the T-slot. A vertical hole through the center of the block mounts the T-bolt and hand knob that lock the stop assembly to the track strip. Another hole drilled lengthwise through the block mounts the flip stop via a carriage bolt with a nylon locknut (a steel nut with a nylon insert that prevents the nut from turning).

The stop itself is cut from $^3/_4$-in. good-quality plywood, such as Baltic or Finnish birch, into an L-shape. A notch on the underside of the stop holds a clear plastic cursor, mounted with a small flat-head screw through a countersunk hole (for instructions on making a cursor, see the sidebar on p. 171). Mark and etch the cross

Eccentric stop offers micro adjustments. Fine adjustments can be made by rotating the stop. The off-center hole makes the position of the stop shift slightly, and the screw locks it down.

hair after the track strip has been installed and calibrated. If you do a lot of dado work or change blades often, additional cross hairs can be added to the cursor to be used with those blades.

To adjust the stop for different-thickness sawblades, you can reposition the track strip, or remove the flip stop from its bolt and add shims (I make these from aluminum beer cans with a leather punch) as necessary. You can also make up different stop assemblies, each with a cursor marked to work with different sawblades, molding heads or dado-blade thicknesses.

Multiple flip stops

Because unused flip stops can be set to desired measurements and then flipped out of the way, several flip stops can be set up along the length of the fence. This would be an advantage if, say, you had to cut all the face-frame components for an entire kitchen to length; stops could be set at all the standard measurements and flipped down whenever needed during cutout. Because flip stops are fairly easy to make, you may wish to make a half-dozen or more at one time. Cut stock for the channels (simple version) or sliding blocks (T-track version) as you would a length of molding; then slice off individual blocks.

Rotating stop handles multiple measurements. This rotating stop allows you to choose one of four stop positions. When used on a drill-press fence, as shown here, it can set distances between closely spaced holes.

Flip stop for mitered ends

Either flip stop described above can be modified to handle boards with mitered ends. If wide picture-frame molding is mitered and the width of a standard stop doesn't catch the tip of the miter, make the face of the stop wider. Alternatively, when making picture frames, it's sometimes desirable to measure distances relative to the inside edge of the frame molding. A shortened flip stop with a 45° triangular block screwed on takes care of this situation, as shown in the drawing on p. 172. A longer cursor must be fitted and etched to register the position the inside edge of the molding butts up to, as shown in the photo on the facing page.

Eccentric end stop

Sometimes you need to position a workpiece along a fence in a fixed position, but in a way that allows some fine-tuning. A simple stop that provides a firm stop, yet provides for a limited amount of adjustment is the eccentric end stop, as shown in the photo on p. 173. I use these as end stops on the pivot arms of my router-plate joinery setup, and they are extremely quick to make. First cut a short length of dowel with a diameter that suits the application. For a small jig, a $1/2$-in.-dia. dowel is about right; for larger jigs, or to yield a greater amount of adjustability, use a 1-in., $1^1/2$-in. or larger diameter dowel. Now drill a hole through the dowel lengthwise that's equidistant between the center and edge. A wood screw through this offset hole mounts the stop to the jig. To make fine adjustments to the stop's position, loosen the screw and rotate the dowel; then lock it in place. You can employ this same principle with even larger stops: Drill an off-center hole in a sawn-out plywood disc, and screw it down where an adjustable stop is needed.

Rotating stop

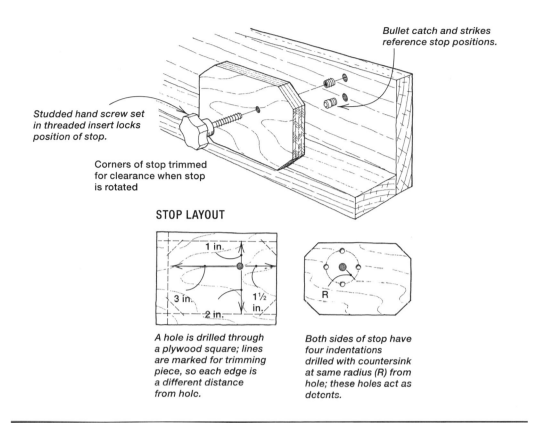

Bullet catch and strikes reference stop positions.

Studded hand screw set in threaded insert locks position of stop.

Corners of stop trimmed for clearance when stop is rotated

STOP LAYOUT

1 in.

3 in.

1½ in.

2 in.

R

A hole is drilled through a plywood square; lines are marked for trimming piece, so each edge is a different distance from hole.

Both sides of stop have four indentations drilled with countersink at same radius (R) from hole; these holes act as detents.

Rotating stop

Sometimes you need to cut, rout or drill two, three or four grooves, shapes or holes that are closely spaced but at a fixed distance from the end of the workpiece. A handy device for this is the rotating end stop, such as the one shown in the photo above. This stop mounts easily to any fence, carriage or table and can be rotated and locked in any of four positions. Each position provides a different spacing between the end of the workpiece and the cutter or bit you are using.

Make the stop by laying out a piece of plywood so that its four sides are each a different distance from a single hole. Start with an oversized piece with a hole marked somewhere in the middle; then use a ruler and a square to mark how the piece must be trimmed (an example is shown in the drawing above). A studded hand knob fits through the hole and into a threaded insert, which is driven into the fence itself.

To allow the fence-mounted stop to clear the jig's base when it is rotated (it's too big diagonally to clear), the corners can be cut off, as on the stop in the photo on the facing page. The position of the rotating stop can be set manually, or detents can be fitted to reference each position.

ABOUT THE AUTHORS

Worth Barton is a design engineer, inventor, and hobbyist woodworker living in San Jose, California.

Joe Beals makes bespoke furniture and architectural millwork in Marshfield, Massachusetts. He is self-taught, about which he makes this comment: "Learning by mistakes is a painful education, but no other method concentrates the attention so perfectly." He also writes fiction and plays traditional Irish music on the fiddle.

Mac Campbell is studying theology in Halifax, Nova Scotia, Canada. Previously, he ran a custom furnituremaking shop in Harvey Station, New Brunswick.

Greg Colegrove is a launch conductor with Lockheed Martin. He supervises countdowns for rockets launching commercial satellites into orbit. He also runs a part-time furnituremaking business in Aurora, Colorado. His specialty is to blend modern hardware and joinery techniques into traditional furniture designs.

Mark Duginske is a woodworker, author, and teacher who offers classes in his shop in Merrill, Wisconsin.

Charles Durham is a professional woodworker in San Clemente, California.

George Fulton is a retired electrical engineer and a hobbyist woodworker. These days he mostly builds furniture for children, which his wife Vivian helps to design. They live in Arnold, Maryland.

Bob Gabor builds Krenov-inspired furniture in Pittsboro, North Carolina. He's also a member of the Triangle Woodworkers Association.

Jeff Greef is a woodworker and journalist living in Santa Cruz, California. He is the author of three books, *Woodshop Jigs and Fixtures, Display Cabinets,* and *Wooden Boxes.* In his spare time, he gardens.

Charles Jacoby is a retired men's clothing store owner who enjoys making furniture for his family in Helena, Montana. Clocks, a portable workbench, and moldings for a 19th-century house are among his current projects. His wife Rosemary helps out.

Tim Hanson is a retired general contractor who used to remodel banks. He now builds furniture and toys in Indianapolis, Indiana.

Jim Hayden is an amateur woodworker and a professional photographer at the Arthur M. Sackler Gallery and Freer Gallery in the Smithsonian Institution, Washington, D.C.

Evan Kern is an author and retired dean of Visual and Performing Arts at Kutztown University. He builds marquetry puzzles inspired by Pennsylvania Dutch art in Kutztown, Pennsylvania.

Edward Koizumi is a professional model maker in Oak Park, Illinois. He builds props for photographers and TV commercials. As a woodworking hobbyist, he works on his Arts and Crafts style house, renovating the kitchen most recently. In his spare time, he rides a tandem bicycle with his son.

William Krase is a retired aerospace engineer at the Rand Corporation who builds furniture and boats in Mendocino, California. He has been a hobbyist woodworker all his life, a hobby that started with model airplanes and has led to 18-ft. sailing boats.

Jeff Kurka mainly builds custom furniture and cabinets although he also designs tools. He does a lot of work for ski areas, having just completed a hall bench for a lodge. An Iowa native, he moved to Florissant, Colorado, five years ago. He also enjoys hunting, fishing, and hiking.

Skip Lauderbaugh is a sales representative for Blum hardware and a college woodworking instructor. His shop is in Costa Mesa, California.

Jim Leslie is an amateur woodturner who dabbles in furnituremaking. He enjoys writing about woodworking and lives in Calgary, Alberta, Canada.

Voicu Marian grew up in his uncle's cabinet shop in Romania. In 1977, he came to the United States and set up shop designing and making furniture part-time. He has been making furniture full-time since 1991. The sawhorses in his article were built after a model in his uncle's shop. He lives in Alliance, Ohio.

Mike McCallum is an artist and inventor who works in product development. He is currently designing a houseboat. He spends his free time with his wife Julia and their daughters. They live between Eugene and Florence, Oregon.

Sandor Nagyszalanczy of Santa Cruz, California, is West Coast Editor of *American Woodworker* magazine and a professional furniture designer and craftsman with more than 20 years of experience. A former senior editor of *Fine Woodworking* magazine, he has written five books including *The Art of Fine Tools*.

Bill Nyberg is director of ophthalmic photography at the University of Pennsylvania in Philadelphia. His yen for woodworking comes from his father, who worked at Dodge Furniture Co. in Manchester, Massachusetts. He designs and makes his own furniture in his spare time.

William Page is a woodworker in Toledo, Ohio.

Guy Perez is a technical writer living in Madison, Wisconsin. In his spare time, he builds woodworking machines to improve his furnituremaking.

Ken Picou is a woodworker and designer of woodworking tools who holds patents on the Robo-Sander and Mr. Mortise tools. He currently builds giant fishing lures (up to 30 in. long) for hopeful fishermen in Austin, Texas.

Gary Rogowski has designed and built fine furniture in Portland, Oregon, since 1974. He runs a school called Northwest Woodworking Studio in Portland. He is a contributing editor to *Fine Woodworking* magazine and the author of *Router Joinery*.

Dale Ross is a professional turner who lives on $5^1/_2$ acres in North Yarmouth, Maine. He runs a nonprofit organization to help and house neglected and abused farm animals. He also enjoys gardening.

Lon Schleining has designed and built stairs in Long Beach, California, for 20 years. He is currently working on a book about curves and bending. He also teaches woodworking at Cerritos College in Norwalk, California.

Jim Siulinski is an applications engineer at Lucent Technologies and runs a small woodshop business on the side in Westbrook, Maine.

Ed Speas works for a Formica corporation as a technical specialist in Henryville, Pennsylvania. He enjoys woodworking as a hobby.

Frank Vucolo owns a staffing business and builds furniture for his home in East Amwell, New Jersey. He enjoys tennis in his spare time.

Pat Warner is a woodworker deep into routing and an instructor at Palomar College in San Marcos, California. He is the author of two books, *Getting the Very Best from Your Router* and *The Router Joinery Handbook*. A third is on its way. He lives in Escondido, California.

Jim Wright is a sheet-metal worker and self-taught woodworker. Recent projects include a wine rack, a sideboard, and a toolbox. He lives in Berkeley, Massachusetts, and his favorite color is aquamarine blue.

PHOTO CREDITS

Jonathan Binzen: p. 152, 153, 154

Mark Duginske: p. 48, 49 (top)

Kent Ezzell: p. 92

Aimé Fraser: p. 16, 20, 115, 116

George Fulton: p. 164

Jeff Greef: p. 145

Boyd Hagen: p. 69 (bottom)

Jim Hayden: p. 5

Charles Jacoby: p. 166

Vincent Laurence: p. 24, 25, 40, 42, 43, 44, 45, 46, 47, 49 (botom), 50, 51, 52, 56, 57, 58, 60, 61, 62, 63, 69 (top), 71, 74, 75, 124, 125, 127, 128, 135, 136, 138

Voicu Marian: p. 19

Mike McCallum: p. 34, 36

Sandor Nagyszalanczy: p. 118, 119, 120, 121, 122, 123, 170, 171, 172, 173, 174

Guy Perez: p. 86, 103 (top)

Strother Purdy: p. 93, 96, 97, 98, 129, 130, 131, 132, 133, 134

Charley Robinson: p. 2, 10, 11, 13, 14, 27, 28, 29, 31, 76, 78, 109

William Sampson: p. 4, 6, 8

Patrick Warner: p. 65, 66, 67, 68

Alec Waters: p. 37, 38, 80, 84, 85, 87, 89, 90, 91, 100, 103 (bottom), 104, 106, 107, 111, 113, 114, 140, 142, 148, 150, 155, 157, 159, 160, 162, 163

EQUIVALENCE CHART

Inches	Centimeters	Millimeters	Inches	Centimeters	Millimeters
$1/8$	0.3	3	12	30.5	305
$1/4$	0.6	6	13	33.0	330
$3/8$	1.0	10	14	35.6	356
$1/2$	1.3	13	15	38.1	381
$5/8$	1.6	16	16	40.6	406
$3/4$	1.9	19	17	43.2	432
$7/8$	2.2	22	18	45.7	457
1	2.5	25	19	48.3	483
$1 1/4$	3.2	32	20	50.8	508
$1 1/2$	3.8	38	21	53.3	533
$1 3/4$	4.4	44	22	55.9	559
2	5.1	51	23	58.4	584
$2 1/2$	6.4	64	24	61.0	610
3	7.6	76	25	63.5	635
$3 1/2$	8.9	89	26	66.0	660
4	10.2	102	27	68.6	686
$4 1/2$	11.4	114	28	71.1	711
5	12.7	127	29	73.7	737
6	15.2	152	30	76.2	762
7	17.8	178	31	78.7	787
8	20.3	203	32	81.3	813
9	22.9	229	33	83.8	838
10	25.4	254	34	86.4	864
11	27.9	279	35	88.9	889
			36	91.4	914

INDEX

INDEX

Publisher: Jim Childs

Associate Publisher: Helen Albert

Associate Editor: Strother Purdy

Copy Editor: Diane Sinitsky

Designer/Layout Artist: Amy Bernard

Indexer: Harriet Hodges

Fine Woodworking magazine

Editor: Timothy D. Schreiner

Art Director: Bob Goodfellow

Managing Editor: Jefferson Kolle

Senior Editors: Jonathan Binzen, Anatole Burkin

Associate Editor: William Duckworth

Assistant Editor: Matthew Teague

Associate Art Director: Michael Pekovich